REED'S
ENGINEERING KNOWLEDGE INSTRUMENTS & CONTROL SYSTEMS FOR DECK OFFICERS

REED'S ENGINEERING KNOWLEDGE INSTRUMENTS & CONTROL SYSTEMS FOR DECK OFFICERS

by

WILLIAM EMBLETON, O.B.E.
Extra First Class Engineer
C.Eng., A.M.I.Mech.E., F.I. Mar. E.

and

THOMAS D. MORTON
Extra First Class Engineer
C.Eng., F.I. Mar. E.

THOMAS REED PUBLICATIONS

A DIVISION OF THE ABR COMPANY LIMITED

First Edition – 1959
Second Edition – 1967
Third Edition – 1973 (SI Units)
Reprinted – 1978
Reprinted – 1980
Fourth Edition – 1985
Fifth Edition – 1995
Reprinted – 2000

ISBN 0 901281 15 8

THOMAS REED PUBLICATIONS
The Barn
Ford Farm
Bradford Leigh
Bradford-on-Avon
Wiltshire BA15 2RP
United Kingdom

E-mail: sales@abreed.demon.co.uk

Produced by Omega Profiles Ltd, SP11 7RW
Printed and Bound in Great Britain

PREFACE

The object of this text book is to present the principles upon which the Department of Transport base the examination papers in Engineering Knowledge, Instruments and Control Systems for Master (foreign-going) and to briefly revise that part of the General Physics for Second Mate which is included in the Master's examination .

Descriptive matter is written in simple language, unnecessary complicated expressions are avoided, and the sketches are in clear diagrammatic form.

Calculations are explained from first principles.

Although the book is intended primarily for Masters, all deck and engineering officers and cadets will find it contains basic and vital principles of engineering so useful to their general marine education.

The authors take this opportunity of acknowledging and thanking the manufacturers mentioned throughout the text for their generosity in supplying information on their products.

In this fifth edition the revision is designed to cover most of the BTEC requirements.

The summary of which aims to provide the Deck Watchkeeping Officer with an understanding of basic engineering and control systems as used on Merchant Vessels and whose pre-requisites are; National Diploma in Maritime Technology and National Diploma in Nautical Science 'A' Level.

W.E.
T.D.M.

CONTENTS

CHAPTER 1

ENGINEERING UNITS AND COMMON TERMS

MASS, FORCE, ENERGY, POWER

Mass is the quantity of matter possessed by a body and is proportional to the volume and the density of the body. It is a constant quantity, that is, the mass of a body can only be changed by adding more matter to it or taking matter away from it.

The abbreviation for mass is *m* and the unit is the kilogramme [kg]. For very large or small quantities, multiples or submultiples of the gramme [g] are used. Large masses are common in marine work and these are measured in megagrammes [Mg]. One megagramme is equal to 10^3 kilogrammes and called a *tonne* [t].

Mass is proportionally accelerated or retarded by an applied force. To maintain a coherent system of units, a unit of force is chosen which will give unit acceleration to unit mass. This unit of force is called the *newton* [N]. Hence, one newton of force acting on one kilogramme of mass will give it an acceleration of one metre per second per second, therefore:

Accelerating force [N] = mass [kg] × acceleration [ms^2]
In symbols:

$$F = ma$$

FORCE OF GRAVITY. All bodies are attracted towards each other, the force of attraction depending upon the masses of the bodies and their distances apart. Newton's law of gravitation states that the force of attraction is proportional to the product of the masses of the bodies and inversely proportional to the square of the distance apart.

An important example of this is the mass of the earth which attracts all comparatively smaller bodies towards it, the attractive force by which a body tends to be drawn towards the centre of the earth is the force of gravity and is called the *weight* of the body.

If a body is allowed to fall freely at the surface of the earth, it will fall with an acceleration of 9.81 m/s^2, this is termed gravitational acceleration and is represented by g. Since one newton is the force which will give one kilogramme of mass an acceleration of one m/s^2, then the force in newtons to give *mkg* of mass an acceleration of 9.81 m/s^2 is $m \times 9.81$. Hence, at the earth's surface, the gravitational force on a mass of m kg is mg newtons, or in other words:

$$\text{weight [N]} = \text{mass [kg]} \times g[\text{m/s}^2]$$

The further the distance between the centre of gravity of the mass and the centre of gravity of the earth, the less is the attractive force between them. Thus, the weight of a mass measured by a spring balance (not a pair of scales which is merely a means of comparing the weight of one mass with another) will vary slightly at different parts of the earth's surface due to the earth not being a perfect sphere.

If a body is projected in a space-rocket, the attractive force of the earth on the body becomes less as its distance from the earth increases until, in complete outer-space, it becomes nil, that is, it is then weightless. The mass of the body of course remains unchanged.

WORK is done when a force applied on a body causes it to move and is measured by the product of the force and the distance through which the force moves.

The unit of work is the *joule* [J] which is defined as the work done when the point of application of a force of one newton moves through a distance of one metre in the direction in which the force is applied. Hence, one joule is equal to one newton-metre. In symbols, J = N m.

$$\text{Work done [J]} = \text{force [N]} \times \text{distance moved [m]}$$

The joule is a small unit. Moderate quantities of work may be expressed in kilojoules [1 kJ = 10^3J] and larger quantities in megajoules [1 Mj = 10^6J].

POWER is the rate of doing work, that is, the quantity of work done in a given time. The unit of power is the *watt* [W] which is equal to the rate of one joule of work being done every second. In symbols, W = J/s = N m/s.

$$\text{Power [W]} = \frac{\text{work done [J]}}{\text{time [s]}} = \text{force [N]} \times \text{velocity [m/s]}$$

The watt is a small unit and only suitable for small powers. For normal powers in engineering, the kilowatt [$1\text{kW} = 10^3\text{W}$] and megawatt [$1\text{MW} = 10^6\text{W}$] are usually more convenient units.

ENERGY is the capacity for doing work and it is measured by the amount of work done. Energy is therefore expressed in the same units as work, that is, joules, kilojoules and megajoules.

Another useful unit of energy is the *kilowatt-hour* [kW h]. This, as its name implies, represents the energy used or the work done when one kilowatt of power is exerted continually for one hour.

$$\begin{aligned} \text{Energy} &= \text{power} \times \text{time} \\ 1\text{kW h} &= 1000 \text{ watts} \times 3600 \text{ seconds} \\ &= 1000\,[\text{J/s}] \times 3600\,[\text{s}] \\ &= 3.6 \times 10^6 \text{ J} \\ &= 3.6 \text{ MJ} \end{aligned}$$

THE MECHANICAL EFFICIENCY is the ratio of the work got out of a machine to the work put into it, and, as this is done in the same time, it is also the ratio of the output power to the input power. Since no machine is perfect, the output is always less than the input, due to frictional and other losses, therefore the efficiency is always less than unity.

The symbol for efficiency is η, the Greek letter eta, and it may be expressed as a fraction or as a percentage.

$$\eta_1 = \frac{\text{output power}}{\text{input power}}$$

PRESSURE

Pressure is expressed as the intensity of force, that is, the force acting on unit area. The unit of force is the newton [N] and the unit of area is the square metre [m^2], therefore the fundamental unit of pressure is the newton per square metre [N/m^2] and this is given the special name *pascal* [Pa]. The symbol representing pressure is usually *p*.

Pressures of fluids in engineering reach high values and these are expressed in multiples of the basic unit of force. For example, the steam pressure in small auxiliary boilers is often in the region of 8×10^5 N/m^2 and in high pressure water-tube boilers it could be 5×10^6 N/m^2. The former can be written 800 kN/m^2 or 800 kPa and the latter 5 MN/m^2 or 5 MPa. Another very convenient unit of pressure commonly used is the bar, which is equal to 10^5 N/m^2. The bar has the advantage of being easy to "think" in these units since one bar is approximately equal to one atmosphere of pressure (1 atm = 1.013 bar), it is also approximately equal to one kilogramme-force per square centimetre which was the unit of pressure used on the Continent before the adoption of SI. Since one bar is 10^5 N/m^2, then the working pressures given in the above examples could be stated as 8 bar in the auxiliary boilers and 50 bar in the water-tube boilers.

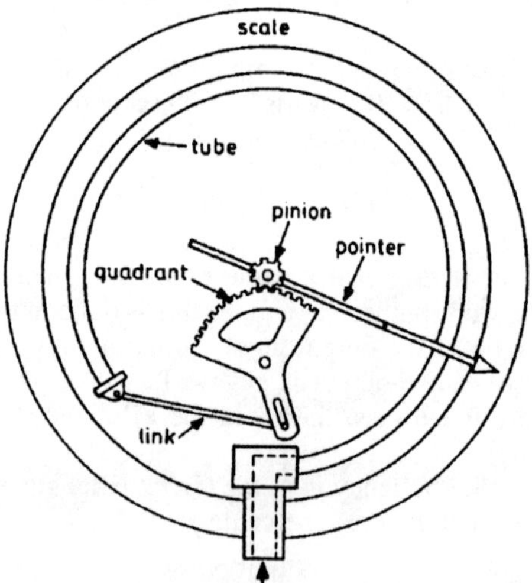

Fig 1
Bourdon Pressure Gauge

PRESSURE GAUGE. The most common instrument used for measuring high static pressures is the Bourdon type of pressure gauge, illustrated in Fig. 1. This consists of a bronze or steel curved tube of elliptical section, one end of the tube is connected to the source of pressure, the other end is sealed. The effect of the fluid pressure inside the tube is to tend to straighten it, the higher the pressure the greater the straightening effect. The small movement of the sealed end of the tube is magnified by linking it to a quadrant meshing with a pinion which carries a pointer on its shaft, the pointer moves over a circular scale on the dial of the gauge which is graduated in bars, kN/m^2, or other suitable units.

MANOMETER. Low pressures and vacua are usually measured in millimetres of mercury [mm Hg], very small pressure such as furnace draught are measured in millimetres of water [mm water]. The instrument used is the manometer which is shown in its simplest form in Fig. 2. This is a glass U-tube partially filled with mercury or water, one end is connected to the source of pressure and the other end is open to the atmosphere. The difference between the levels of the liquid in the two legs indicate the difference in pressure between the source and the atmosphere.

Considering the manometer containing mercury, if we take the density of mercury as 13.6×10^3 kg/m^3 and the force of gravity on a mass of one kilogramme as 9.806 65 newtons (a more accurate figure for the standard value of gravitational acceleration than 9.81 which is usually acceptable in engineering), then the weight of one cubic metre of mercury is $13.6 \times 10^3 \times 9.806\ 65$ newtons = 133.3 kN. Hence a column of mercury one metre high exerts a pressure of 133.3 kN on one square metre, or a column of mercury one millimetre high is equivalent to a pressure of 133.3 N/m^2.

Similarly, each millimetre of water pressure is equal to 9.806 65 N/m^2 which is usually taken as 9.81 N/m^2.

Small pressures may also be expressed in millibars [mbar]

One mbar = $1 \text{ bar} \times 10^{-3} = 10^5 \times 10^{-3}$ N/m^2 = 100N/m^2

BAROMETER. The mercurial barometer works on the principle of the atmospheric pressure supporting a column of mercury. The simplest is constructed with a glass tube about 800mm long, closed at one end, open at the other. The tube is completely filled with mercury so that all air is excluded and, with the open end plugged, the tube is inverted and its open end submerged into a vessel containing mercury. When the plug is removed, the level of the mercury falls, leaving a perfect vacuum between the mercury level and the sealed end (see Fig. 3). The vertical column of mercury left standing up the tube is supported by the outside atmospheric pressure and is therefore a measure of the pressure of the atmosphere. As the atmospheric pressure rises and falls, the level of

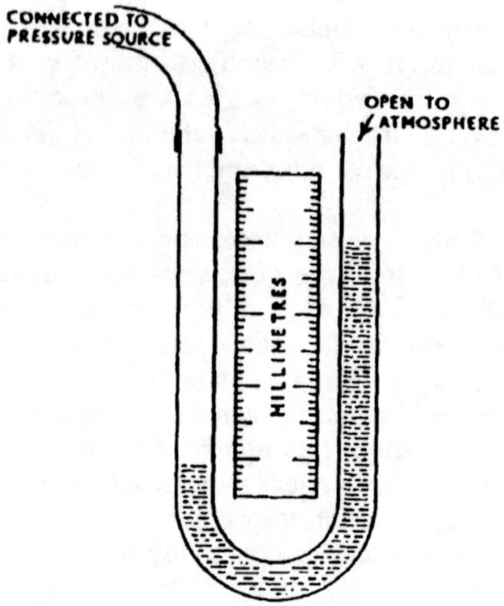
CONNECTED TO
PRESSURE SOURCE
OPEN TO
ATMOSPHERE
MILLIMETRES

Fig 2

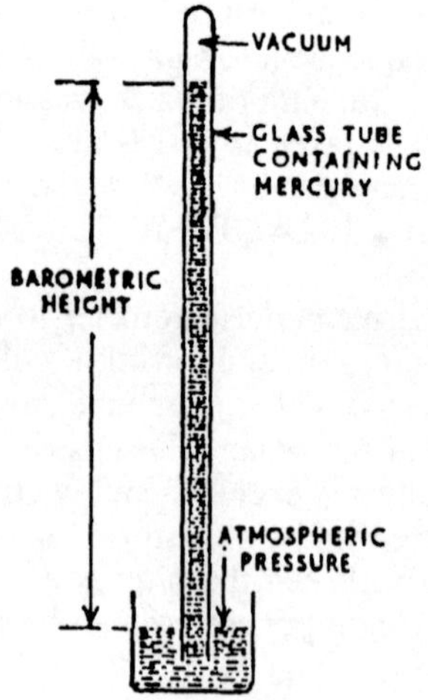
VACUUM
GLASS TUBE
CONTAINING
MERCURY
BAROMETRIC
HEIGHT
ATMOSPHERIC
PRESSURE

Fig 3

the supported column of mercury rises and falls accordingly.

For example, if the column of mercury supported by the atmospheric pressure is 760mm, then the atmospheric pressure will be:

$$760 \times 133.3 = 1.013 \times 10^5 \text{ N/m}^2 = 1.013 \text{ bar}$$

GAUGE PRESSURE AND ABSOLUTE PRESSURE. Most pressure recording instruments, including the ordinary pressure gauge and the open ended manometer, measure the pressure from the level of atmospheric pressure. The pressure so recorded is termed the gauge pressure and the word 'gauge' should follow the units of pressure. Thus, if a steam pressure gauge reads 20 bar the pressure should be stated as 20 bar gauge, meaning that this is the pressure of the steam over and above the atmospheric pressure.

The true pressure is measured above a perfect vacuum and called the *absolute pressure* and this is the value which is used in thermodynamic calculations. The absolute pressure is therefore obtained by adding the atmospheric pressure to the gauge pressure, the gauge pressure being read from the pressure gauge and the atmospheric pressure obtained from the barometric reading.

As an example, if the pressure of a fluid is 5.5 bar gauge, and the barometer stands at 758mm Hg then,

$$\begin{aligned} \text{atmospheric pressure} &= 758 \times 133.3 \\ &= 1.01 \times 10^5 \text{ N/m}^2 = 1.01 \text{ bar} \\ \text{absolute pressure} &= \text{gauge pressure} + \text{atmospheric pressure} \\ &= 5.5 + 1.01 \\ &= 6.51 \text{ bar} \end{aligned}$$

This could be written 6.51 bar absolute. However it is usual to omit the word absolute and take it for granted that if the word gauge does not follow the value of the pressure then it means that it is an absolute pressure.

Pressure gauges are not always perfectly accurate and, in any case, it is difficult to read to an accuracy of one or two kN/m^2. It is therefore quite common when exact accuracy is not essential to assume the atmospheric pressure to be 100 kN/m^2, which is 1 bar.

In the above example, if the barometer reading was not known, the absolute pressure would be taken as:

$$5.5 + 1 = 6.5 \text{ bar}$$

with very little difference in the final result of a calculation.

VACUUM GAUGE. If the manometer shown in Fig. 2 is used as a vacuum gauge, say for a steam condenser, the level of the mercury in the leg connected to the condenser will be higher than the level in the leg open to atmosphere. The difference in level indicates the pressure *below* atmospheric and written 'mm Hg vacuum.'

For example, if the gauge reads 600mm Hg of vacuum and the barometer stands at 758mm Hg then:

Pressure below atmospheric
$$= 600 \times 133.3 = 8 \times 10^4 \text{ N/m}^2 = 800 \text{ millibars}$$
Atmospheric pressure
$$= 758 \times 133.3 = 1.01 \times 10^5 \text{ N/m}^2 = 1{,}010 \text{ millibars.}$$

Therefore absolute pressure in condenser is 800mbar below 1,010mbar, which is 210mbar, or more simply calculated:

$$\begin{aligned} \text{abs. press.} &= (\text{barometer mm Hg} - \text{vac. gauge mm Hg}) \times 133.3 \\ &= (758 - 600) \times 133.3 \\ &= 158 \times 133.3 = 2.1 \times 10^4 \text{ N/m}^2 \\ &= 210 \text{ mbar} \end{aligned}$$

VOLUME

The basic unit of volume is the cubic metre [m^3] and the symbol for volume is *V*. A common submultiple is the *litre* [l], this is equal in volume to one cubic decimetre and is used for fluid measure.

$$1 \text{ m}^3 = 10^3 \text{ dm}^3 \text{ therefore } 10^3 \text{ litres} = 1\text{m}^3$$

The millilitre [ml] is 1×10^{-3} litre and therefore equal in volume to one cubic centimetre. Whereas the basic unit of density is kilogramme per cubic metre [kg/m^3], densities of liquids are often expressed in grammes per millilitre [g/ml] and densities of solids in grammes per cubic centimetre [g/cm^3].

SPECIFIC VOLUME is the volume occupied by unit mass, the symbol is *v* and the basic unit is cubic metre per kilogramme [m3/kg], thus the specific volume is the reciprocal of density. In certain cases, specific volume may be expressed in cubic metres per tonne [m^3/tonne] and litres per kilogramme [l/kg].

TEMPERATURE

Temperature is an indication of hotness or coldness and therefore is a measure of the intensity of heat.

The most common temperature measuring instrument is the mercurial thermometer This consists of a glass tube of very fine bore with a bulb at its lower end, the bulb and tube are exhausted of air, partially filled with mercury and hermetically sealed at the top end. When the thermometer is placed in a substance whose temperature is to be measured, the mercury takes up the same temperature and expands (if heated) or contracts (if cooled) and the level, which rises or falls in consequence indicates on the thermometer scale the degree of heat intensity

The divisions on the thermometer are called degrees and are related to the freezing and boiling points of pure water at atmospheric pressure.

The Celsius scale (formerly known as Centigrade) is the SI system of measuring and specifying temperatures. The point at which pure water freezes into ice is marked zero, and the point at which pure water boils into steam at atmospheric pressure is assigned the number 100. The former is sometimes referred to as the *lower fixed point* or *ice point*, the latter as the *upper fixed point* or *steam point*. The number of degrees into which the thermometer scale is divided between these two fixed points is therefore 100, and the sizes of these degrees are continued uniformly above and below the two fixed points for measuring higher and lower temperatures.

The unit representing a temperature reading is °C, and the symbol for temperature is θ, the Greek letter theta. The difference between two temperature readings is referred to as a *temperature interval.*

ABSOLUTE TEMPERATURE. All gases expand at practically the same rate when heated through the same range of temperature, and contract at the same rate when cooled.

The rate of expansion or contraction of a perfect gas is (very nearly) $^1/_{273}$ of its volume at 0°C when heated or cooled at constant pressure through one degree Celsius. Hence, if a gas initially at 0°C could be cooled at constant pressure until its temperature is 273 degrees below 0°C, the volume would contract until there was nothing left and no further reduction of temperature would be possible, that is, the gas would then have reached its *absolute zero of temperature* (see Fig. 4). In practice of course, it is not possible to

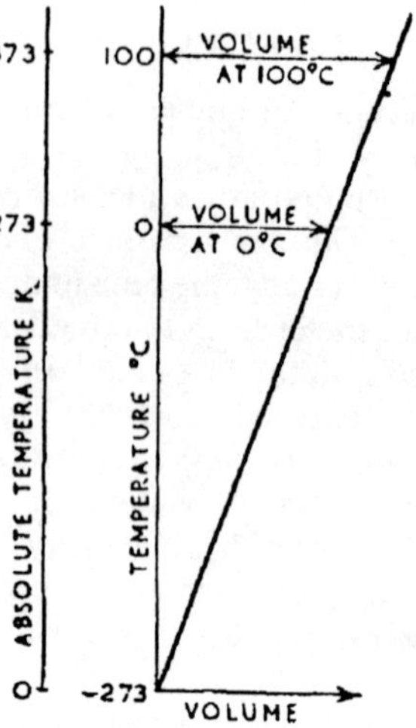

Fig 4

cool a gas down to the absolute zero and cause it to disappear. As the absolute zero of temperature is approached the gas will change into a liquid and the laws of gases are then no longer applicable.

We see from the above that temperatures can be expressed as absolute quantities, that is, stating the degrees of temperature above the level of Absolute Zero, by adding 273 to the ordinary Celsius thermometer reading.

Absolute temperature is often referred to as *thermodynamic temperature*, the symbol for this is T and the unit is the kelvin which is represented by K, thus,

thermodynamic temperature = Celsius temperature + 273

In symbols,

$$T[\text{K}] = \theta[^\circ\text{C}] + 273$$

HEAT

HEAT is a form of energy associated with the movement of the molecules which constitute the heated body. It is interchangeable with other forms of energy and can be made available for doing work and producing mechanical and electrical power.

The basic unit of all energy, including heat, is the *joule* [J]. Thus, units of heat are expressed in joules or multiples of the joule, the most common being kilojoules [kJ] and megajoules [MJ].

THE SPECIFIC HEAT of a substance is the quantity of heat required to raise the temperature of unit mass of the substance by one degree. Hence, the total quantity of heat energy transferred to a substance to raise its temperature is the product of the mass of the substance, its specific heat, and its rise in temperature. Different substances have different specific heat values.

LATENT HEAT is the heat which supplies the energy necessary to overcome some of the binding forces of attraction between the molecules of a substance and is responsible for it changing its physical state from a solid into a liquid, or from a liquid into a vapour, the change taking place without any change of temperature.

The process of changing the physical state from a solid into a liquid is called *melting* or *fusion,* and the quantity of heat required to change unit mass of the substance from solid to liquid at the same temperature is the *latent heat of fusion.*

The process of changing the physical state of a substance from a liquid into a vapour is called *boiling* or *evaporation* and the quantity of heat to bring about this change at constant temperature to unit mass is the *latent heat of evaporation.*

The latent heat of evaporation of water at atmospheric pressure is 2,256.7kJ/kg. This means that one kilogramme of water at 100°C would require 2,256.7 kilojoules of heat to completely boil it into one kilogramme of steam at 100°C. Also one kilogramme of steam at 100°C would require to lose 2,256.7 kilojoules of heat to completely condense it into one kilogramme of water at 100°C.

The temperature at which a liquid boils, and the latent heat of evaporation, depend strictly upon the pressure, the higher the pressure the higher the boiling point and the smaller the amount of latent heat.

PROPERTIES OF STEAM

SATURATED STEAM is steam which is in physical contact with the boiling water from which it was generated, its temperature is the same as the boiling water and this is referred to as the *saturation temperature.* If the vapour produced is pure steam at this temperature, it is called *dry saturated steam.* If the steam contains water (usually very fine particles held in suspension in the form of a mist) it is called *wet saturated steam,* or sometimes more briefly as *wet steam.*

DRYNESS FRACTION. The quality of wet steam is expressed by its dryness fraction, which is the ratio of the mass of pure steam in a given mass of the steam-plus-water mixture. For example, if 2kg of "steam" is taken directly from the steam space of a boiler, tested and found that this was composed of 0.06kg of water and 1.94kg of pure steam, its dryness fraction is $1.94 \div 2 = 0.97$.

SUPERHEATED STEAM. In order to increase the temperature of steam above its saturation temperature without increasing the pressure, the steam must be taken away from its contact with the water from which it was generated and heated externally as it passes to the engines. Steam whose temperature is higher than its saturation temperature corresponding to its pressure is termed *superheated steam*. When steam is superheated, its volume increases approximately in proportion to its increase in absolute temperature.

As superheated steam is at a higher temperature and greater volume than saturated steam at the same pressure, more heat energy is stored in each kilogramme. This extra energy can produce more power in the engines and the efficiency is thereby increased.

Whereas saturated steam begins to condense immediately it comes into contact with engine parts of a lower temperature than the steam, superheated steam contains heat above its saturation temperature and therefore initial condensation and power loss due to this, is reduced.

Hence the main advantages of using superheated steam are, for the same power, the engines consume less steam and therefore less fuel is required. This results in smaller boiler capacity and less fuel to be carried for a given voyage. There is also less likelihood of water hammer in steam pipes, and less initial impact loss and erosion of turbine blades because of the absence of water moisture.

Against the above advantages, superheated steam is not so suitable as saturated steam for reciprocating machinery as there are no water particles to act as a lubricant on shuttle and slide valve faces and piston rings. A desuperheater is usually fitted in the main steam range for supplying steam to the auxiliaries. Better quality metal is required for valves and seats, turbine blades and nozzles, and any parts in which superheated steam comes into contact. The metal must retain most of its strength at high temperatures and resist erosion. Monel metal and stainless steels are most suitable.

CHAPTER TWO

PROPERTIES OF GASES

When a substance has been evaporated it can exist as a gas or vapour and one of its most important characteristics is its elastic property. For instance, if a certain volume of a liquid is put into a vessel of larger volume, the liquid will only partially fill the vessel, taking up no more nor less volume than it did before, but when a gas enters a vessel it immediately fills up every part of that vessel no matter how large it is. Practically speaking, liquids cannot be compressed nor expanded, but gases can be compressed into smaller volumes and expanded to larger volumes.

A perfect gas is a theoretically ideal gas which strictly follows Boyle's and Charles' laws of gases.

Consider a given mass of a perfect gas enclosed in a cylinder by a gas-tight movable piston. When the piston is pushed inward, the gas is compressed to a smaller volume, when pulled outward the gas is expanded to a larger volume. However, not only is there a change in volume but the pressure and temperature also change. These three quantities, pressure, volume and temperature, are related to each other, and to determine their relationship it is usual to perform experiments with each one of these quantities in turn kept constant while observing the relationship between the other two.

In such basic laws, the pressure and temperature must be the absolute values, that is, measured from absolute zero, and not measured from some artificial level. Absolute values were explained in Chapter 1.

BOYLE'S LAW

Boyle's law states that the absolute pressure of a fixed mass of a perfect gas varies inversely as its volume if the temperature remains unchanged.

$$p \propto \frac{1}{V} \text{ therefore } p \times V = \text{a constant.}$$

Hence, $$p_1 \times V_1 = p_2 \times V_2$$

To illustrate this, imagine 2m³ of gas at a pressure of 100kN/m² (= 10⁵N/m² = 1 bar) contained in a cylinder with a gas-tight movable piston as illustrated in Fig. 5. When the piston is pushed inward the pressure will increase as the gas is compressed to a smaller volume and, provided the temperature remains unchanged, the product of pressure and volume will be a constant quantity for all positions of the piston. From the known initial conditions the constant is calculated:

$$p_1 \times V_1 = \text{constant}$$
$$100 \times 2 = 200$$

and the pressure at any other volume can be determined:

When the volume is 1.5m^3,
$$p_2 \times 1.5 = 200$$
$$p_2 = 133.3\text{kN/m}^2$$
When the volume is 1m^3,
$$p_3 \times 1 = 200$$
$$p_3 = 200\text{kN/m}^2$$

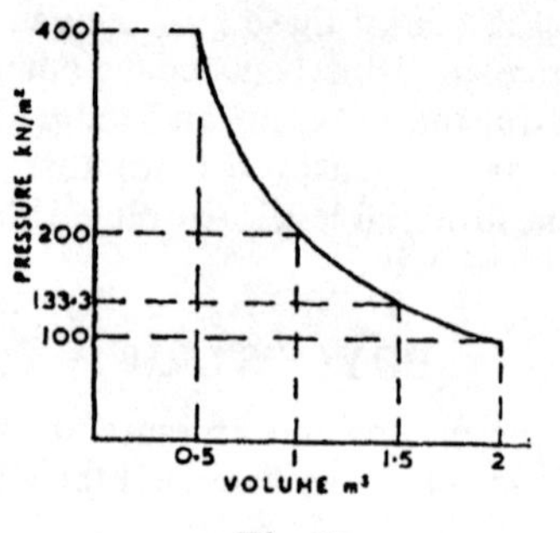

Fig 5

When the volume is $0.5m^3$,

$$p_4 \times 0.5 = 200$$
$$p_4 = 400kN/m^2$$

And so on.

The variation of pressure with the change of volume is shown in the graph below the cylinder in Fig. 5. The graph produced by joining up the plotted points is a regular hyperbola, consequently we refer to compression or expansion where pV = constant as hyperbolic compression or hyperbolic expansion. When the temperature is constant as in this example, the operation may also be termed "isothermal."

Note that as the ordinates (vertical measurements) represent pressure, and the abscissae (horizontal measurements) represent volume, and since the product of pressure and volume is constant, then all rectangles drawn from the axes with their corners touching the curve, will be of equal area.

Example, $3.5m^3$ of air at a pressure of $20kN/m^2$ gauge is compressed at constant temperature to a pressure of $425kN/m^2$ gauge. Taking the atmospheric pressure as $100kN/m^2$ calculate the final volume of the air.

Initial absolute pressure = 20 + 100 = 120 kN/m^2
Final absolute pressure = 425 + 100 = 525 kN/m^2

$$p_1V_1 = p_2V_2$$
$$120 \times 3.5 = 525 \times V_2$$
$$V_2 = \frac{120 \times 3.5}{525} = 0.8m^3 \qquad \text{Ans.}$$

CHARLES' LAW

Charles' law states that the volume of a fixed mass of a perfect gas varies directly as its absolute temperature if the pressure remains unchanged, also, the absolute pressure varies directly as the absolute temperature if the volume remains unchanged.

From the above statement we have:

For constant pressure, $V \alpha T$ $\qquad \therefore \frac{V}{T} = \text{constant}$

hence, $\frac{V_1}{T_1} = \frac{V_2}{T_2}$ $\qquad$ or $\frac{V_1}{V_2} = \frac{T_1}{T_2}$

For constant volume, $p \propto T$ $\qquad \therefore \dfrac{p}{T} = \text{constant}$

hence, $\dfrac{P_1}{T_1} = \dfrac{P_2}{T_2}$ $\qquad$ or $\dfrac{P_1}{P_2} = \dfrac{T_1}{T_2}$

Example. The pressure of the air in a starting air vessel is 40 bar (= $40 \times 10^5 \text{N/m}^2$) and the temperature is 24°C. If a fire in the vicinity causes the temperature to rise to 65°C, find the pressure of the air. Neglect any increase in volume of the vessel.

As the term "gauge" does not follow the given pressure, it is assumed that this is the initial absolute pressure.

Initial absolute temperature = 24°C + 273 = 297K
Final absolute temperature = 65°C + 273 = 338K

$$\frac{p_1}{T_1} = \frac{P_2}{T_2} \qquad \therefore \quad p_2 = \frac{p_1 T_2}{T_1}$$

$$p_2 = \frac{40 \times 338}{297} = 45.52\text{bar} \qquad \text{Ans.}$$

COMBINATION OF BOYLE'S AND CHARLES' LAWS

Each one of these laws states how one quantity varies with another if the third quantity remains unchanged, but if the three quantities change simultaneously, it is necessary to combine these laws in

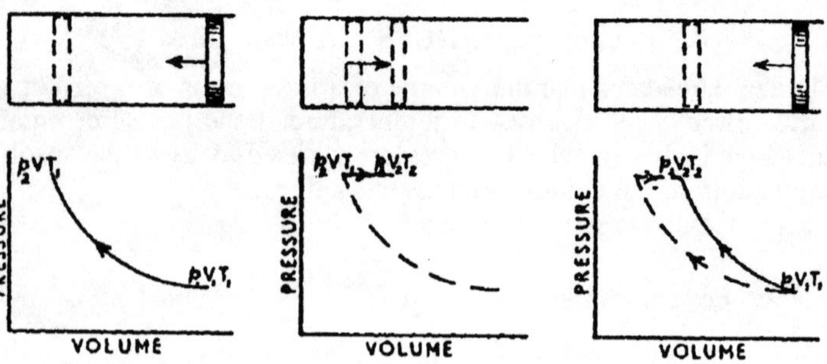

Fig 6

order to determine the final conditions of the gas.

Referring to Fig. 6 which again represents a cylinder with a piston, gas-tight so that the mass of gas within the cylinder is always the same. Let the gas be compressed from its initial state of pressure p_1, volume V_1, and temperature T_1, to its final state of p_2V_2 and T_2, but to arrive at the final state let it pass through two stages, the first to satisfy Boyle's law and the second to satisfy Charles' law.

Imagine the piston pushed inward to compress the gas until it reaches the final pressure of p_2 and let its volume then be represented by V. Normally the temperature would tend to increase due to the work done in compressing the gas, but any heat so generated must be taken away from it during compression so that its temperature remains unchanged at T, hence following Boyle's law:

$$p_1V_1 = p_2V \quad \text{.......................................} (i)$$

Now apply heat to raise the temperature from T_1 to T_2 and at the same time draw the piston outward to prevent a rise of pressure and keep it constant at p_2. The volume will increase in direct proportion to the increase in absolute temperature according to Charles' law:

$$\frac{V_2}{V} = \frac{T_2}{T_1} \quad \text{.......................................} (ii)$$

By substituting the value of V from (ii) into (i) this quantity will be eliminated:

From (ii) $$V = \frac{V_2T_1}{T_2}$$

Substituting into (i)

$$p_1V_1 = p_2 \times \frac{V_2T_1}{T_2}$$

$$\therefore \frac{p_1V_1}{T_1} = \frac{p_2V_2}{T_2}$$

This combined law of Boyle's and Charles' is true for a given mass of any perfect gas subjected to any form of compression or expansion.

Example. 0.5m³ of a perfect gas at a pressure of 0.95 bar and temperature 17°C are compressed to a volume of 0.125m³ and the final pressure is 5.6 bar. Calculate the final temperature.

Initial absolute temperature = 17°C + 273 = 290K

$$\frac{p_1 V_1}{T_1} = \frac{p_2 V_2}{T_2}$$

$$\frac{0.95 \times 0.5}{290} = \frac{5.6 \times 0.125}{T_2}$$

$$T_2 = \frac{5.6 \times 0.125 \times 290}{0.95 \times 0.5}$$

$$= 427.4\text{K}$$

427.4 – 273 = 154.4°C Ans.

Example. The volume of a certain mass of gas is 500cm³ when its pressure is 1,026 millibars and temperature 27°C. Calculate its volume at 1,013 millibars and 0°C.

$$\frac{p_1 V_1}{T_1} = \frac{p_2 V_2}{T_2}$$

$$\frac{1026 \times 500}{(27 + 273)} = \frac{1013 \times V_2}{(0 + 273)}$$

$$V_2 = \frac{1026 \times 500 \times 273}{1013 \times 300}$$

$$= 461 \text{ cm}^3$$ Ans

COMPRESSION OF A GAS IN A CLOSED SYSTEM

When a gas is compressed in a cylinder by the inward movement of a gas-tight piston (Fig. 7), the pressure of the gas increases as the volume decreases. The work done *on* the gas to compress it appears as heat energy in the gas and the temperature tends to rise. This effect can readily be seen with a tyre inflator; in pumping up the

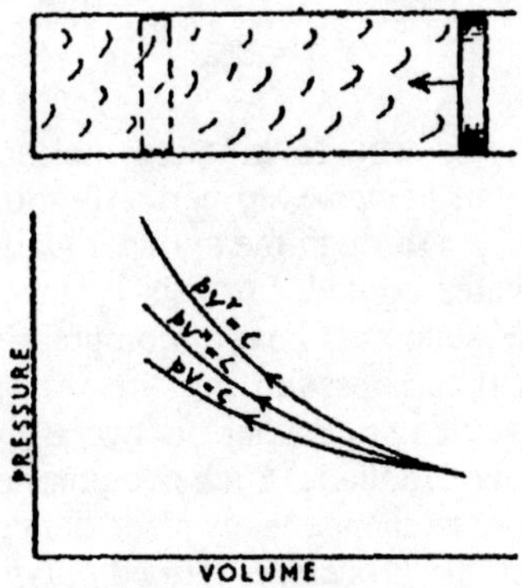

Fig 7

tyre the discharge end of the inflator gets hot due to compressing the air.

ISOTHERMAL COMPRESSION. Imagine the piston pushed inward slowly to compress the gas and, at the same time, let heat be taken away via the cylinder walls (by a water-jacket or other means) to avoid any rise in temperature. If the gas could be compressed in this manner, *at constant temperature*, the process would referred to as *isothermal compression* and the relationship between pressure volume would follow Boyle's law:

$$pV = \text{constant} \qquad \therefore p_1V_1 = p_2V_2$$

ADIABATIC COMPRESSION. Now imagine the piston pushed inward quickly so that there is insufficient time for any heat energy to be transferred from the gas to the cylinder walls. All the work done in compressing the gas appears as stored up heat energy. The temperature at the end of compression will therefore be high and, for the same ratio of compression as the first case, the pressure will consequently be higher. This form of compression, where no heat energy transfer takes place between the gas and an external source, is known as *adiabatic compression*. The relationship between pressure and volume for adiabatic compression is:

$$pV^\gamma = \text{constant} \qquad \therefore p_1V_1^\gamma = p_2V_2^\gamma$$

where γ (gamma) is the ratio of the specific heat of the gas at constant pressure to the specific heat at constant volume, thus,

$$\gamma = \frac{C_p}{C_v}$$

POLYTROPIC COMPRESSION. In practice, neither isothermal nor adiabatic processes can be achieved perfectly. Some heat energy is always lost from the gas through the cylinder walls, more especially if the cylinder is water cooled, but this is never as much as the whole amount of the generated heat of compression. Consequently, the compression curve representing the relationship between pressure and volume lies somewhere between the two theoretical cases of isothermal and adiabatic. Such compression, where a partial amount of heat energy exchange takes place between the gas and an outside source during the process, is termed *polytropic compression* and the compression curve follows the law:

$$pV^n = \text{constant} \qquad \therefore p_1V_1^n = p_2V_2^n$$

Thus the law pV^n = constant may be taken as the general case to cover all forms of compression from isothermal to adiabatic wherein the value of n for isothermal compression is unity, for adiabatic compression $n = \gamma$ and for polytropic compression n generally lies somewhere between I and γ.

EXPANSION OF A GAS IN A CLOSED SYSTEM

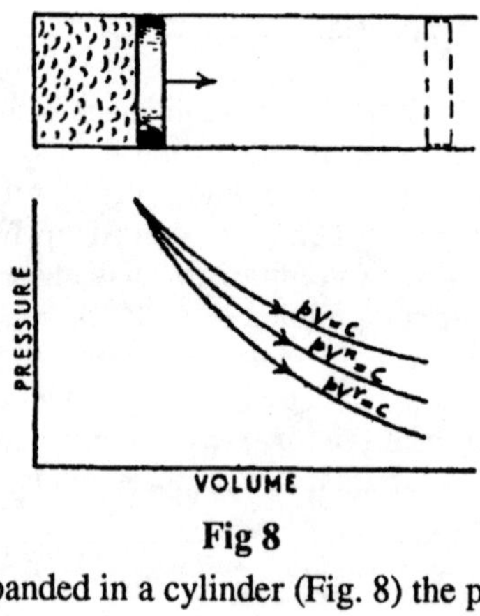

Fig 8

When a gas is expanded in a cylinder (Fig. 8) the pressure falls and the volume increases as the piston is pushed outward by the energy in the gas.

This is exactly the opposite to compression. Work is done *by* the gas in pushing the piston outward and there is a tendency for the temperature to fall due to the heat energy in the gas being converted into mechanical energy. Therefore to expand the gas *isothermally*, heat energy must be transferred to the gas from an external source during the expansion in order to maintain its temperature constant. The expansion would then follow Boyle's law, $pV = \text{constant}$.

The gas would expand *adiabatically* if no heat energy transfer, to or from the gas, occurs during expansion, the external work done in pushing the piston forward being entirely at the expense of the stored up heat energy. Therefore the temperature of the gas will fall during the expansion. As for adiabatic compression, the law for adiabatic expansion is $pV^{\gamma} = \text{constant}$.

During a *polytropic* expansion, a partial amount of heat energy will be transferred to the gas from an outside source but not sufficient to maintain a uniform temperature during the expansion. The law for polytropic expansion is $pV^n = \text{constant}$ as it is for polytropic compression.

With reference to Figs. 7 and 8 the student should note that the adiabatic curve is the steepest, the isothermal curve is the least steep, and the polytropic curve lies between the two. Thus, the higher the index of the law of expansion or compression, the steeper will be the curve.

It must also be noted that for any mode of expansion or compression in a closed system, the combination of Boyle's and Charles' laws is always true:

$$\frac{pV}{T} = \text{constant} \qquad \therefore \frac{p_1V_1}{T_1} = \frac{p_2V_2}{T_2}$$

Example. 0.25m³ of air at 90kN/m² and 10°C are compressed in an engine cylinder to a volume of 0.05m³, the law of compression being $pV^{1.4} = \text{constant}$. Calculate (i) the final pressure, (ii) the final temperature.

$$p_1V_1^{1.4} = p_2V_2^{1.4}$$

$$90 \times 0.25^{1.4} = p_2 \times 0.05^{1.4}$$

$$p_2 = \frac{90 \times 0.25^{1.4}}{0.05^{1.4}} = 90 \times \left\{\frac{0.25}{0.05}\right\}^{1.4}$$

$$= 90 \times 5^{1.4} = 856.7\text{kN/m}^2 \qquad \text{Ans. (i)}$$

$$\frac{p_1V_1}{T_1} = \frac{p_2V_2}{T_2}$$

$$\frac{90 \times 0.25}{(10 + 273)} = \frac{856.7 \times 0.05}{T_2}$$

$$T_2 = \frac{283 \times 856\text{-}7 \times 0.05}{90 \times 0.25} = 538.8 \text{ K}$$

$$538.8 - 273 = 265.8°\text{C} \qquad \text{Ans. (ii)}$$

Example. 0.07m³ of gas at 41.4 bar is expanded in an engine cylinder and the pressure at the end of expansion is 3.1 bar. If expansion follows the law $pV^{1.35}$ = constant, find the final volume.

$$p_1V_1^{1.35} = p_2V_2^{1.35}$$

$$41.4 \times 0.07^{1.35} = 3.1 \times V_2^{1.35}$$

$$V_2 1.35 = \frac{41.4 \times 0.07^{1.35}}{3.1}$$

$$V_2 = 0.07 \times \sqrt[1.35]{\frac{41.4}{3.1}} = 0.477\text{m}^3 \text{ Ans.}$$

Example. 0.014m³ of gas at 31.5 bar is expanded in a closed system to a volume of 0.154m³ and the final pressure is 1.2 bar. If the expansion takes place according to the law pV^n = constant, find the value of n.

$$p_1V_1^n = p_2V_2^n$$

$$31.5 \times 0.014^n = 1.2 \times 0.154^n$$

$$\frac{31.5}{1.2} = \left\{\frac{0.154}{0.014}\right\}^n$$

$$26.25 = 11^n$$

$$\log 26.25 = n \times \log 11$$

$$1.4191 = n \times 1.0414$$

$$n = \frac{1.4191}{1.0414} = 1.363 \qquad \text{Ans.}$$

CHAPTER THREE

MARINE BOILERS

The reader should revise Heat and Properties of Steam in Chapter 1 before proceeding to the thermal process of steam generation in a Marine Boiler to be described. The temperature at which water changes into steam, that is the boiling point of water, depends strictly upon the pressure exerted on it. A few examples are: –

Pressure (bar)	0.13	1	10	20	100
Boiling point (°C)	50	100	180	212	311

That is ranging from a low pressure evaporator operating at a vacuum bar to tank type boilers 10, to 20 bar then up to a water tube boiler at 100 bar.

If a boiler is working at a pressure of 100 bar the water begins to boil when its temperature reaches 311°C and the steam is generated at the same temperature. This saturated steam flows away from the water from which it was made and is subjected to further heating which increases the dryness of the steam but does not increase its temperature.

If the steam flow passes out of the boiler to externally heated elements its temperature can be increased and the steam becomes "superheated" before passing to the steam engine. In the engine the steam loses its superheat, expands, does work and finally condenses into water which is returned to the boiler as feed water to be once again converted into steam and so the cycle is complete.

There are two distinct types of marine boiler in use on board ship in the Merchant Navy, the fire-tube boiler in which the hot gases from the furnaces pass through the tubes while the water is on the outside, and the water-tube boiler in which the water flows through the inside of the tubes while the hot furnace gases pass around the outside.

Scotch boilers were commonly used in conjunction with steam

reciprocating engines but because of their comparative poor efficiency and low power/weight ratio, these systems are no longer installed. Scotch boilers may be found in service on older vessels as auxiliary units.

DONKEY BOILER

A donkey boiler is an auxiliary boiler sometimes included in a steamship's installation for supplying steam to the auxiliaries in port when the main boilers are shut down, and in motorships with steam auxiliaries for supplying steam at sea and in port.

A cross sectional sketch of the Cochran Spheroid boiler is shown in Fig. 9. This auxiliary boiler has an all welded shell, a seamless spherically shaped furnace and small bore tubes which are expanded at their ends into flat tube plates. Advantages claimed over earlier designs are:

1) Increased steam output for the same size as earlier designs.
2) Spherical furnace gives increased radiant heating surface and is the ideal shape for withstanding pressure
3) Efficient (up to 80%) and robust.

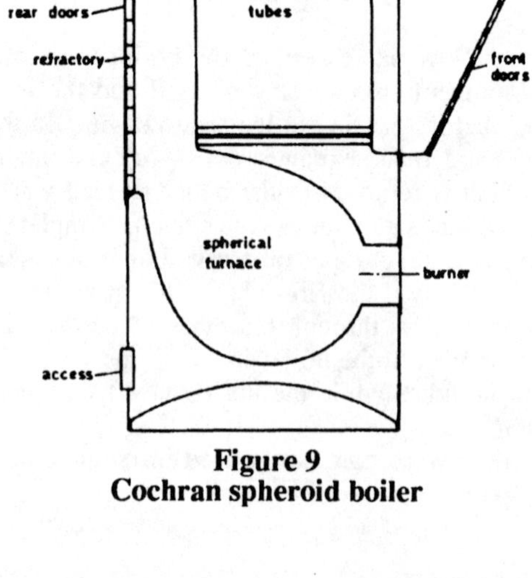

Figure 9
Cochran spheroid boiler

4) Easier to maintain.
5) No furnace brickwork required apart from burner quarls.
6) With small tubes, fitted with retarders, gas velocity and turbulence are increased. This gives better heat transfer and cleaner tubes.

It can be supplied in a range of sizes from 1.5 to 2.6m diameter with height just over twice the diameter and working pressure 17 to 10 bar.

Most modern boilers are automatic. A simple way of understanding automatic operation is to mentally go through the manual process required to produce and control steam flow from a boiler. Viz for lighting up: 1. Water in boiler and at correct level. 2. Air to burners and furnace for combustion and purging. 3. Fuel supply adequate and at correct pressure and temperature. 4. Means of ignition. 5. Light the burner. Eventually steam flow at the correct pressure is sensed and the signal is used to vary the firing rate as the demand requires. Boiler water level must also be sensed and water supply varied according to demand. Fig. 10 shows a simplified control system for a boiler.

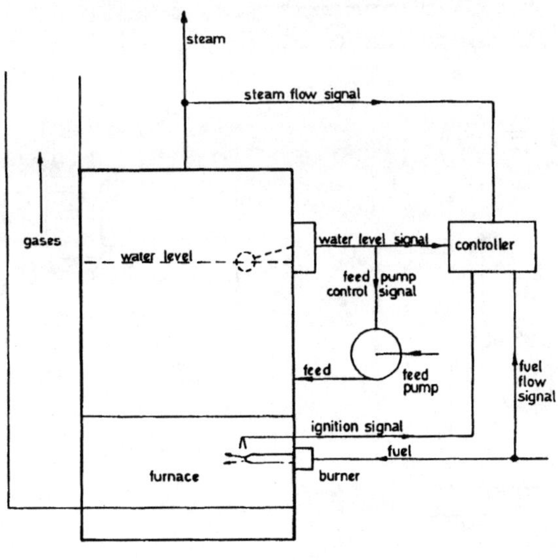

Figure 10
Boiler control

EXHAUST GAS AND COMPOSITE BOILERS

In motor ships, the heat carried away in the exhaust gases from the main engines amounts to between 20 and 30% of the total heat given out by the fuel burnt in the cylinders. This is a heavy loss and, to recover some of this heat which would otherwise be wasted, the exhaust gases may be passed through an exhaust gas steam boiler (sometimes called a waste-heat boiler) before allowing it to escape up the funnel. The steam thus generated drives auxiliaries.

The amount of heat extracted from the exhaust gases depends upon the quantity and temperature of the exhaust gas and the

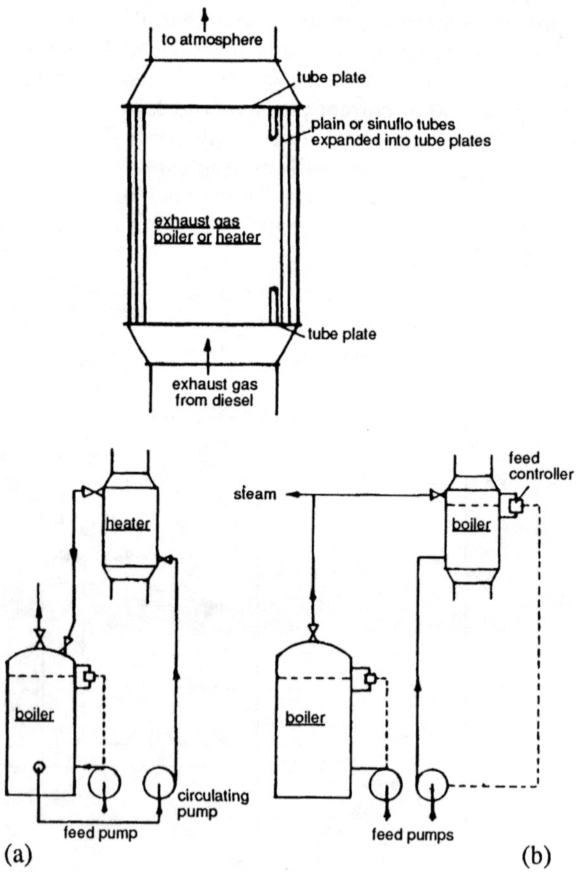

Figure 11

temperature (and pressure) of the steam generated, however, it is sometimes possible to recover as much as 40% of the heat in the exhaust gases.

A modern Cochran exhaust gas boiler consists of all welded tube and wrapper plates made of good quality boiler steel. Tubes, made of electric resistance welded mild steel, are swelled at one end then expanded into the tube plates. This type of boiler would be fitted in the exhaust gas uptake from the diesel engine. It may be connected to a Cochran Spheroid boiler as shown in Fig. 11(a) in which case it would be operating full as a water heater. Fig. 11(b) shows a flexible arrangement whereby either or both the boilers could be in operation depending upon requirements.

Composite boilers are often fitted wherein the generation of steam can be maintained by oil firing when the exhaust gas temperature falls due to slow running of the engines, or ceases when the engines are stopped.

Fig. 12 shows a composite Cochran boiler. Separate uptakes are provided for the engine exhaust and the products of combustion from the oil fire.

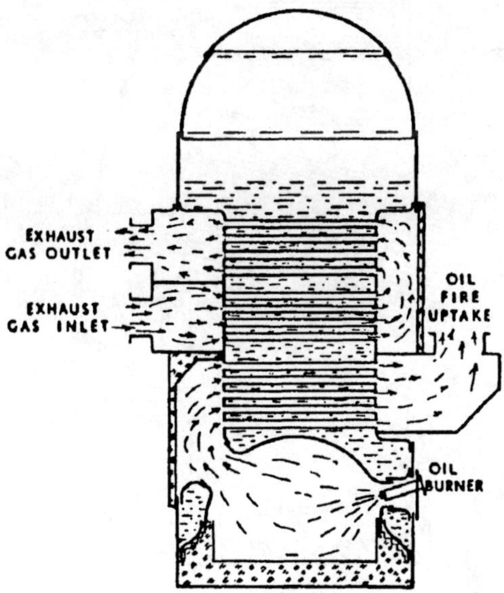

Figure 12
Composite Cochran boiler

WATER TUBE BOILERS

Water tube boilers produce large quantities of high pressure superheated steam for use in steam turbine machinery. Some smaller water tube boilers, such as the header type Fig. 13 producing lower pressure steam, may be used as an auxiliary boiler. Steam from the higher pressure boilers is also used after suitable economic temperature and pressure reduction coupled with, in some cases, superheat loss. Used in auxiliary machinery *e.g.* turbo-generators, turbo-feed pumps, reciprocating pumps, heaters etc.

There are many types of water-tube boilers in general use on board ship, such as the Babcock and Wilcox, Yarrow and Foster Wheeler, and each manufacturer has a variety of designs to offer depending upon requirements. Water-tube boilers have distinct advantages over fire tube boilers; (i) circulation of the water is natural and immediate upon lighting up, thus steam can be raised from cold water in a matter of a few hours without the danger experienced in fire tube boilers of temperature differences between

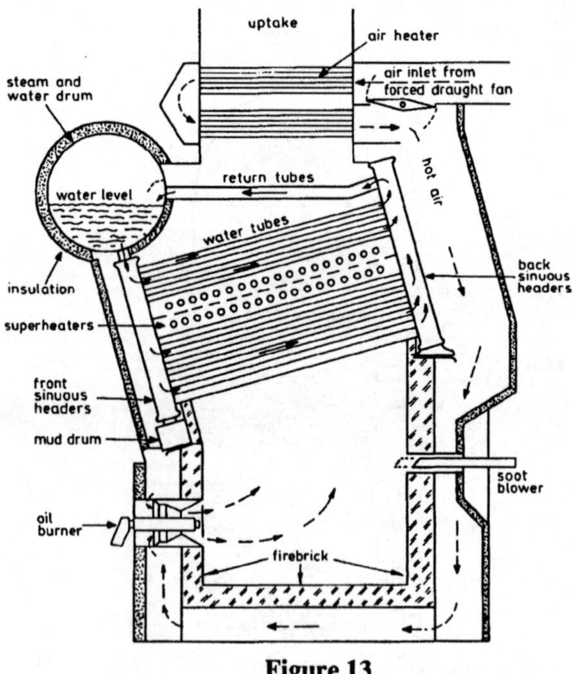

Figure 13
Babcock and Wilcox water tube boiler

top and bottom parts of the boiler causing unequal expansion resulting in mechanical straining; (ii) the steam and water drums are small in diameter compared with *e.g.*, the large shell of the fire tube boiler, therefore they are stronger and suitable for much higher steam pressures; (iii) less mass of water is carried in the boiler, hence a saving in weight; (iv) less space is taken up for the same output.

The water-tube boiler does however require greater skill in operating and maintaining, especially with regard to the purity of the feed water, and its upkeep is more costly than the fire tube boiler.

BABCOCK AND WILCOX BOILER. One type is shown diagrammatically in Fig. 13. This is the header type of boiler and consists of a steam and water drum, front headers and back headers which are connected by tubes inclined at about 15 degrees. The headers are boxes of square plan section, are sinuous to enable successive rows of tubes to be staggered, and have small handholes provided with doors opposite the tube ends to permit access to the tubes for cleaning, tube replacement, expanding etc. The path of the water as it circulates in the boiler is shown by full lines and arrows from the drum, into the front headers, through the tubes, into the back headers and through the return tubes back into the drum. Sometimes the hot gases and flames from the fires are baffled to direct them to take a tortuous path before escaping to the uptake.

The superheater is made up of a number of horizontal U-tubes placed transversely across the boiler, either immediately above the water tubes or in between the banks of tubes as shown in Fig. 13. The inlet and outlet ends of the U-tubes are expanded into the saturated and superheated steam headers respectively which run along one side of the boiler.

A FOSTER WHEELER "D" TYPE WATER-TUBE BOILER is shown in Fig. 14. This consists of two horizontal cylindrical drums, one above the other, the top being the steam and water drum (referred to briefly as the steam drum) and the bottom the water drum, these are connected directly by vertical generating tubes, and by other tubes via headers. One set of tubes from the steam drum are bent at approximately the middle of their length so that the upper portion forms part of the roof of the combustion chamber and the lower portion forms the side wall, these tubes are connected to a side header at combustion chamber floor level which, in turn, is connected to the water drum by a bottom set of tubes to form part of the floor of the combustion chamber. Other sets of tubes, by being bent at right angles, connect the steam drum to an upper rear header to form another roof layer, vertical tubes from upper rear header to lower rear header form a back wall, and finally more tubes with right angle bends connect the lower rear header to the water drum to form another layer over the floor of the combustion chamber.

The superheater consists of U-tubes expanded into vertical headers located between the two banks of generating tubes. Saturated steam is led by a pipe from the top of the steam drum to the bottom of the superheater inlet header, and the superheated steam leaves the outlet header at the top of the superheater.

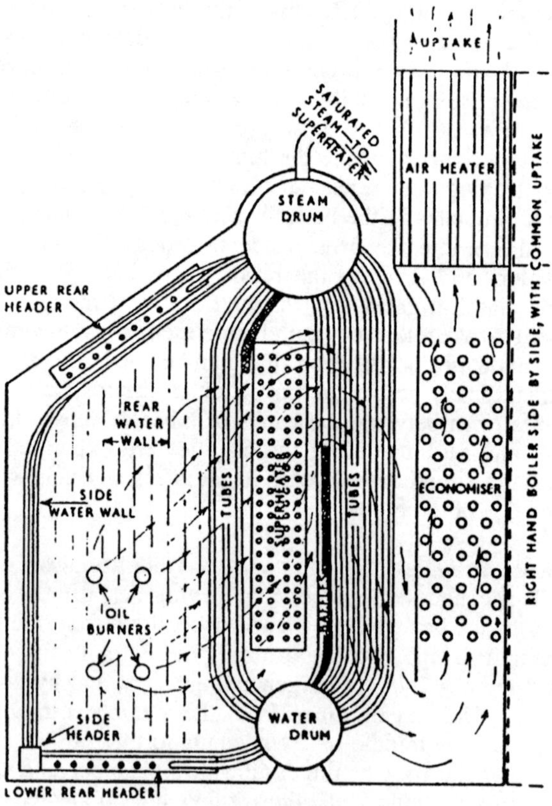

Figure 14
Foster Wheeler D type boiler

The water circulates upwards through the vertical generating tubes nearest the oil burners and downwards through the vertical bank of generating tubes behind the superheater. Further water circulation takes place from the water drum along the floor tubes and up through the wall tubes to the steam drum, then down from steam drum to water drum via the generating tubes behind the superheater.

Baffles are fitted to direct the hot gases from the oil burners upwards over the nearest bank of generating tubes and superheater, downwards through the further main bank of generating tubes to the bottom of the economiser, then up through economiser and air-heater to the uptake and funnel.

The economiser is so called because economy is effected by transferring heat from the hot waste gases to the boiler feed water. This heat which would otherwise be lost in the gases escaping up the funnel, increases the temperature of the feed water, hence less heat is required from the fuel per given mass of steam generated in the boiler, resulting in increased efficiency. The economiser consists of a bank of horizontal steel tubes in staggered formation, sometimes with external gills shrunk on to increase the surface area and assist transference of heat.

The whole system of drums and tubes is encased in sheet steel with fire-brickwork bolted to the inside of the casing. The boilers are usually fitted in pairs so that one common uptake can be arranged for each pair of boilers.

WATER TUBE BOILERS WITH MEMBRANE WALLS

Membrane walls consist of many small diameter water tubes with strips of steel welded between them to form a panel. The panels can be shaped to form the shell of a water tube boiler. Floor, roof and walls are made up of membrane sections whose tube ends are attached to headers and drums to form a tall gas tight box. A vertical screen membrane directly connected to steam and water drums divides the boiler internally, separating the furnace from the gas uptake. This screen has the tubes staggered at its lower end to form a path for the flue gases. A similar arrangement exists in the roof for burner penetration and the flue gas outlet.

Fig. 15 shows a roof fired membrane wall boiler for pressures up to 110 bar and up to 226,000kg/h steam output (11,000kw). Considerable advantages are claimed, 1. No refractory required. 2. Gas tight casing hence no gases can enter the boiler room. 3. Reduced cost. 4. High efficiency.

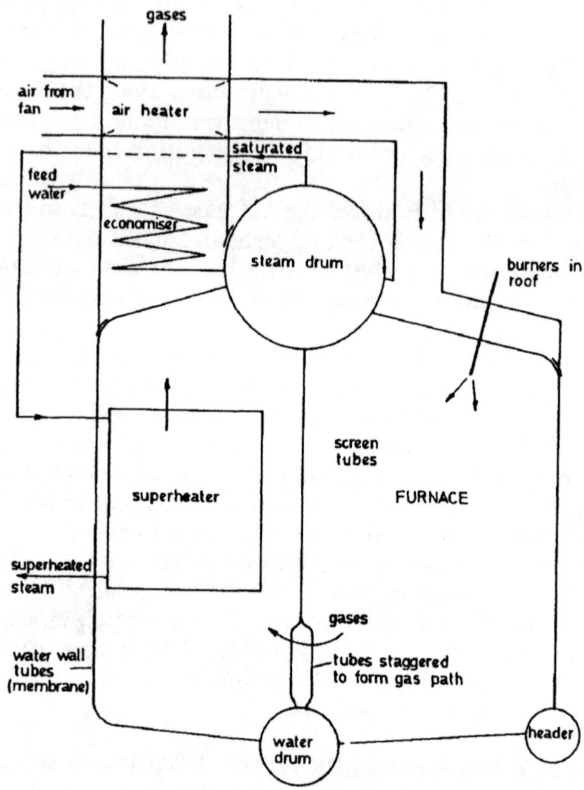

Figure 15
Water tube boiler

BOILER MOUNTINGS

Boiler mountings are the valves and cocks fitted to the boiler, the usual fittings are listed below:

MAIN STEAM STOP VALVE, fitted on top of the boiler and connected to the main steam pipe line which carries the steam to the main engines. This is a screw-down valve and may or may not be a non-return type. Its function is to isolate the boiler from the main steam line and therefore it is either full open when the boiler is supplying steam to the main engines, or tight closed when the boiler is not supplying main steam.

AUXILIARY STEAM STOP VALVE, fitted on top of at least one of the boilers and connected to the auxiliary steam pipe line which carries steam to the various steam pumps, steam engine driven electric generators, winches, etc. It is of the screw-down non-return type.

MAIN FEED CHECK VALVE connected to the main feed water pipe line from the feed pumps. It is a screw-down non-return type of valve and is operated by the engineer on watch to regulate the amount of water fed into the boiler. Its position is on the shell of the boiler, either at water level (with an extended spindle and handwheel down to a convenient height for operation from the stokehold platform) where it leads directly to a spray pipe inside the boiler at water level; or the valve may be positioned lower down on the shell, in which case an internal pipe will lead the water up to the spray pipe.

AUXILIARY FEED CHECK VALVE connected to the auxiliary feed water pipe line from the feed pumps. It is a similar valve to the main feed check and is positioned alongside it. It is principally for use in port when the auxiliaries are running and the main engines shut down but can be used as an alternative to the main feed if anything should go wrong with that line.

FEED WATER REGULATOR. This is a feed controller which automatically maintains the correct water-level and is essential for water-tube boilers with high evaporative rates. Such a fitting relieves the engine-room personnel of hand regulation of the feed check valve, maintains steadier steaming conditions and reduces the risks of water shortage and priming.

Due to the small quantity of water contained and the high evaporative rate of high pressure water tube boilers their water level is controlled automatically.

If during manœuvring an increase in steam flow caused by opening further the steam supply valve to the turbine occurs this would lead to a reduction in boiler pressure. A rise in water level (swell effect) would take place and the feed valve would be automatically opened to *increase* feed water flow, this despite the fact that the water level has risen! It is necessary in order to meet the increase in steam demand.

The water level will gradually return to its desired value and a new balanced boiler throughput exists.

(For low rated boilers, when the water level rises the feed valve would be closed in to reduce the water supply. This would not be satisfactory in highly rated boilers.)

When steam demand is reduced by closing in the steam supply valve the boiler pressure increases. Boiler fuel supply, and hence heat supply, is reduced and the water level will fall (shrink effect). In low rated boilers the feed supply would be increased to restore water level but in highly rated boilers the opposite must occur since if the steam demand were to suddenly increase the swell effect might cause priming, water hammer and damage.

WATER HAMMER. Water hammer is the impact of water upon steam pipes and fittings. It can be caused by priming a boiler so that water is carried over into the piping and turbines. It could strip the blades from the turbines, it could shatter valves or fracture piping with disastrous results.

It can also be due to incorrect or no drainage of condensed steam in the system and the steam being suddenly turned on. If water is lying in a steam pipe and the steam supply is rapidly turned on the water is pushed along the pipe, before it has time to evaporate, at considerable speed and when it reaches a discontinuity *e.g.* pipe bend or valve its impact may result in damage.

Care must be taken to drain and warm through steam piping systems gradually and carefully. Initially ,with the drains open , just crack open the steam supply valve and do not open it further until steam is flowing freely out of the drains. If steam passes over water in the piping and the water is moved to form a wave that completely fills the pipe the steam ahead of the wave could be condensing and reducing pressure.This coupled with the steam pushing the wave along can increase the speed of the water further and hence increase the impact effect.

WATER GAUGES. The water gauge is a glass tube or plates gripped in steam tight glands with steam and water shut-off cocks, thus the water level in the boiler can be seen as it takes up the same level in the glass. The steam cock is fitted in the steam space of the boiler, the water cock in the water space, and the gauge is so positioned that the glass is half-full of water when the boiler water is at working level. It is most essential to maintain the correct water level. If the water level falls too low there is danger of the top rows of tubes becoming exposed due to shortage of water, these parts would then become overheated, lose strength and distort or collapse under pressure. If the water-level is allowed to rise too high there is danger of water being carried over with the steam into the engines (this is known as "priming") with resultant damage to the engines.

For small boilers and on the drums of water-tube boilers, the

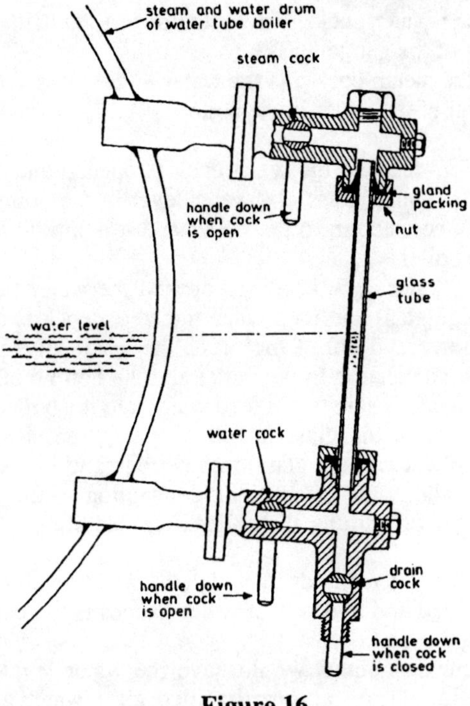

Figure 16
Water gauge

water gauge is fitted directly on the shell (Fig. 16). This is the simplest arrangement and has three cocks only, a steam cock, a water cock, and a drain cock, the working conditions being steam and water cocks open and drain cock shut. The water gauge should be blown through at regular intervals, usually once a watch, to prevent accumulation of foreign matter which might choke the passages and result in a false water reading. To blow this gauge:

(1) Close steam and water cocks then open the drain. Nothing should then blow out of the gauge if the steam and water cocks are not leaking.
(2) Open and close water cock to check that the water cock

connection to the boiler is clear.
(3) Open and close steam cock to check that the steam cock connection to the boiler is clear.
(4) Close the drain.
(5) Open the water cock. Water should then gradually rise up to the top of the gauge glass.
(6) Open the steam cock and the water in the glass should fall to the level of the water in the boiler.

If when (5) is reached the water cock is opened and water does not flow up the gauge glass, the water level in the boiler is below the water cock connection to the boiler and it is unsafe to put feed water into the boiler.

If when the water cock has been opened the water flows to the top of the gauge glass and then when the steam cock is opened the water flows down and out of the glass, the water level is between the water cock connection to the boiler and the bottom of the gauge glass. In this case it is safe to put feed water into the boiler.

If after (5) when the glass is full of water, the steam cock is opened and the water in the glass does not descend in the glass, the water level is above the steam cock connection to the boiler and there is a danger of priming the boiler if any additional feed is put into it.

There must be two methods by which the water level in the boiler can be ascertained and two sets of water gauges is the usual way to satisfy this regulation.

Most present day plants would have the water level indicated remotely in addition to the directly fitted glass water gauge. The remote display, together with high and low water level alarms, would be situated in the engine room control room or in some convenient place in the engine room. A duplicate could also be found in the bridge control room.

PRESSURE GAUGE COCK in the steam space to which is coupled the connecting pipe to the steam pressure gauge at convenient eye level. The Bourdon type of pressure gauge was explained in Chapter 1.

SAFETY VALVES. Spring loaded safety valves are mounted high up in the steam space of the boiler. When the steam exceeds a pre-determined pressure the valve lifts against the compression of a helical spring and allows the steam to escape through an outlet branch in the valve chest which is connected to the waste steam pipe leading up or alongside the funnel to atmosphere. At least two valves must be fitted so that if one became defective the other could deal with any excess steam being generated. They usually constitute a pair in a common valve chest. Spring loaded valves are always the type fitted on ships' boilers because the blow-off pressure, which depends upon the adjustment of the compression of the spring, is constant and not affected by the rolling of the ship.

Spring loaded valves have the disadvantage of accumulation of pressure, that is, as the valves are lifting, more pressure is necessary to compress the spring and lift them further. The valves are set to blow off at a pressure not exceeding 3% of the working pressure and must be so designed that, under full firing conditions of the boiler and with the steam stop valve shut, the valves must lift sufficiently so that the pressure does not rise above 10% of the designed blow-off pressure. Specially shaped lips on valve and seat assist in minimising accumulation of pressure. The safety valves are fitted with hand-operated easing gear by which the valves can be manually opened in case of emergency.

HIGH LIFT SAFETY VALVES. An improvement on the ordinary spring loaded safety valve is the *high-lift* type. This is, as its name implies, designed to give a higher lift than normal to allow a greater area of escape for the steam.

In the Cockburn improved high lift safety valve, a piston is fitted

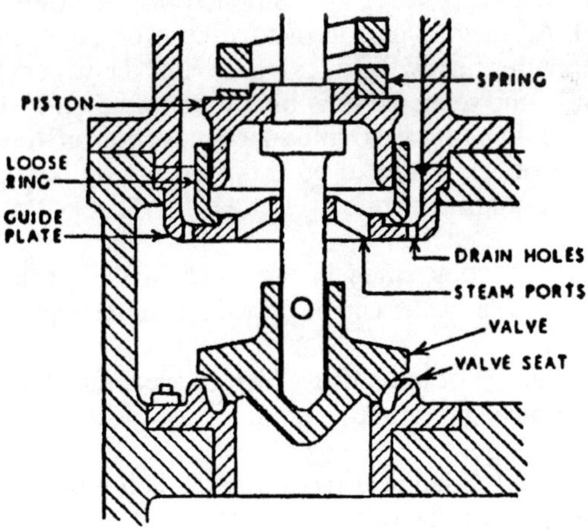

Figure 17
High lift safety valve

to the spindle at the upper part of the waste steam space of the valve chest. This piston, which is larger in diameter than the valve, works inside a loose ring within a guide plate (see Fig. 17). The guide plate centralises the spindle and has ports through to allow the passage of waste steam to the underside of the piston, and thus a further additional effort acts on the valve spindle by means of the waste steam pressure on the piston. The valve itself is the normal type with lip as in ordinary safety valves but has no wings, it is centralised on its seat by being a fairly good fit on the valve spindle.

CIRCULATING VALVE at the bottom of fire tube boilers, connected by a pipe to the suction side of the feed pumps through which the water is drawn to circulate when raising steam.

BLOW DOWN VALVE at the bottom of the boiler, connected by a pipe to the blow down cock on the ship's side. This valve (and ship's side cock) is opened when it is required to empty the boiler in port prior to opening up for cleaning and inspection when time does not permit the slower but safer procedure of allowing the water to completely cool down and then pump it overboard.

Sometimes there is an extra branch on the blow down valve chest of fire tube boilers connected to the circulating pipe, instead of having a separate circulating valve.

SCUM VALVE. This valve is positioned on the boiler just below water level. An internal pipe leads to a scoop or pan open at the top, a few centimetres below the working level of the water. The outside connection of the scum valve is by a pipe to the blow down ship's side cock. Its function is to allow accumulation of foreign matter which floats on the water to be skimmed off by blowing the top layer of water out of the boiler.

SALINOMETER COCK fitted in the water space through which a sample of the boiler water can be drawn off for testing.

AIR COCK on top of the boiler which is opened to allow the air to escape when raising steam from cold.

WHISTLE VALVE fitted near the top of the boiler and connected by a pipe to the steam whistle.

SOOT BLOWERS must be fitted in fire tube boilers which have

superheaters inside the smoke tubes, as brush cleaning of the tubes is not possible. They are always fitted to water-tube boilers, to clean the spaces between the water tubes, economisers and superheaters.

Maintaining boiler heating surfaces clean, both on the gas and water sides keeps boiler efficiency high and fuel consumption down.

There are different types of soot blowers. Fig. 18 shows diagrammatically the principle of a typical hand-operated steam-jet blower. This is a tube with angled steam nozzles of heat resisting steel in the end, which can be rotated and also projected towards or retracted from the tubes. The movement of the hand-wheel turns the screw working inside a nut in the tube, this causes lateral movement of the tube and at the same time, by means of a fixed pin projecting into a scroll in the nut, the tube acquires the desired rotary movement. The lateral movement of the tube also opens and closes the steam ports. Thus, when the steam valve is opened to the blower, the one operation of turning the handwheel controls the steam admission and movement of the nozzles to direct fine high pressure jets of steam in the required direction and move by a sweeping action over the tube area. When not in use, the nozzle end of the blower is retracted inside its housing as a protection against damage due to overheating and burning.

A number of blowers may be fitted at different points within a boiler to ensure complete effective coverage, and blowing is performed at regular intervals, perhaps once every 12 hours depending upon steaming conditions, to maintain clean surfaces.

Another type of soot blower uses compressed air instead of

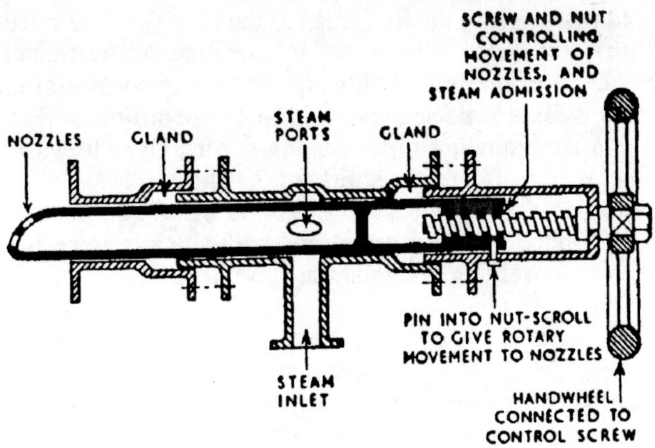

Figure 18
Soot blower

steam. This requires the extra installation of an air compressor but has the advantage of reducing the quantity of boiler feed water to be made up due to steam losses. The air jet type is also more suitable for mechanically controlled automatic sequence operation.

COMBUSTION

Combustion of the fuel in the furnaces of boilers and in the cylinders of internal combustion engines is the chemical combination of the combustible elements in the fuel with the oxygen of air. The principal combustibles in all fuels are carbon and hydrogen. Average oil fuel requires a theoretical minimum of about 14kg of air to completely burn 1kg of oil. The air must be intimately mixed with the fuel and the amount of fuel which can be burned depends upon the quantity of air supplied. An excess of the theoretical minimum quantity of air is always necessary, the amount of excess depending upon the design of the combustion spaces and conditions under which the fuel is burned. If an attempt is made to burn fuel with an insufficient air supply, combustion will not be complete, the first indication of this being black smoke from the funnel; if too much air is supplied an unnecessary amount of heat will be carried away to waste up the funnel; each case represents a loss of efficiency.

The calorific value (heating value) of oil is about 42 to 45 MJ/kg.

FURNACE DRAUGHT. Natural draught of air to boiler furnaces is caused by the hot gases rising up the funnel and air from the stokehold rushing into the furnaces to take its place. The intensity of natural draught depends upon the temperature of the funnel gases and the height of the funnel, thus, to obtain a good natural draught, these two evils are necessary; a high temperature of flue gases results in a large amount of heat being carried away to waste and a tall funnel would affect the stability of the ship.

Assisted draught systems are therefore installed to enable more air to be supplied to the furnaces, and consequently more fuel to be burned, without relying upon natural conditions.

FORCED DRAUGHT. A common forced draught system consists of a large fan in the engine room, steam engine or electric motor driven, which delivers air along an air duct to an enclosed furnace front. Included in the air ducting range is a tubular air-heater in the boiler uptake. The furnace front is a closed box with air valves to control the air supply.

The advantages of a forced draught system over natural draught are: better control of air supply therefore greater control of power; the air is heated prior to admission to the furnaces and this heat being reclaimed from the flue gases means that less heat of the fuel is wasted in heating the furnace air up to combustion temperature and not so much cooling effect on the furnaces; more air can be supplied therefore more fuel can be burned to generate more steam, hence smaller or fewer boilers are required.

Another method of producing more draught is the Induced Draught System. This consists of a large fan installed in the uptake which pulls the gases through the boiler and pushes them up the funnel, thereby inducing more air into the furnaces.

The intensity of the draught is measured by a hydrostatic gauge (Fig. 2) which consists of a glass U-tube containing water, one leg being connected to the fan casing and the other leg open to atmosphere. The difference in water level expresses the draught pressure in millimetres of water, each millimetre being equal to 9.81 N/m^2.

AIR PREHEATERS are fitted in the boiler uptake. The commonest form is a system of tubes over or through which the air required for combustion passes on its way to the furnace, the hot uptake gases on the other side of the tube walls heat the tubes and consequently pre-heats the air.

The advantages of preheating are that heat which would be otherwise lost is extracted from the flue gases before escaping up the funnel, and therefore less heat goes to waste. Hot instead of cold air is supplied to the furnace, therefore there is less mechanical straining of the boiler parts due to difference in temperatures, less heat is lost in bringing the air up to combustion temperature and the efficiency of the boiler is thereby increased.

OIL BURNING

A diagrammatic sketch of an oil burning system is shown in Fig. 19. The oil fuel is stored in the double-bottom tanks and pumped from there as required into settling tanks by means of a transfer pump. The settling tanks are situated in the stokehold, two tanks each having a capacity of about 12 hours supply are usually installed so that the oil can stand for about 12 hours to allow any water in the oil to settle to the bottom and be drained off before use. The oil is drawn from the settling tank, through cold filters into the fuel pressure pump which discharges it at a pressure from 5.5 bar

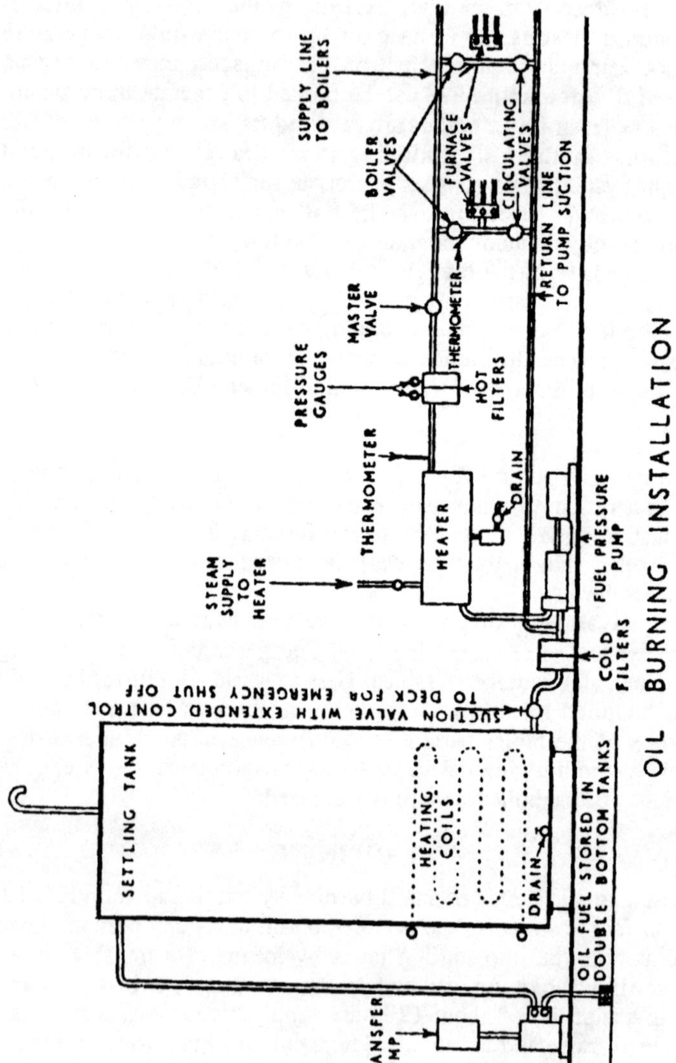
OIL BURNING INSTALLATION
SUPPLY LINE TO BOILERS
BOILER VALVES
FURNACE VALVES
CIRCULATING VALVES
RETURN LINE TO PUMP SUCTION
MASTER VALVE
THERMOMETER
PRESSURE GAUGES
HOT FILTERS
THERMOMETER
DRAIN
HEATER
STEAM SUPPLY TO HEATER
FUEL PRESSURE PUMP
COLD FILTERS
SUCTION VALVE WITH EXTENDED CONTROL TO DECK FOR EMERGENCY SHUT OFF
SETTLING TANK
HEATING COILS
DRAIN
OIL FUEL STORED IN DOUBLE BOTTOM TANKS
TRANSFER PUMP

Figure 19

upwards (depending upon the design of the system) through a steam operated heater where it is heated to about 90°C (depending upon the class of oil), through hot filters into the supply line to the sprayers in the boiler furnaces.

There is an oil circulating valve on the furnace distribution valve chest of every boiler which allows the oil to be returned to the pump suction so that any cold oil lying in the pipe line is cleared out for hot oil to take its place before attempting to light a fire.

The transfer pump, fuel pressure pump, heater and filters are in duplicate with cross-connections so that either set can be used while the other set is a stand-by.

The burners (Fig. 20) sometimes called sprayers, are fitted in the

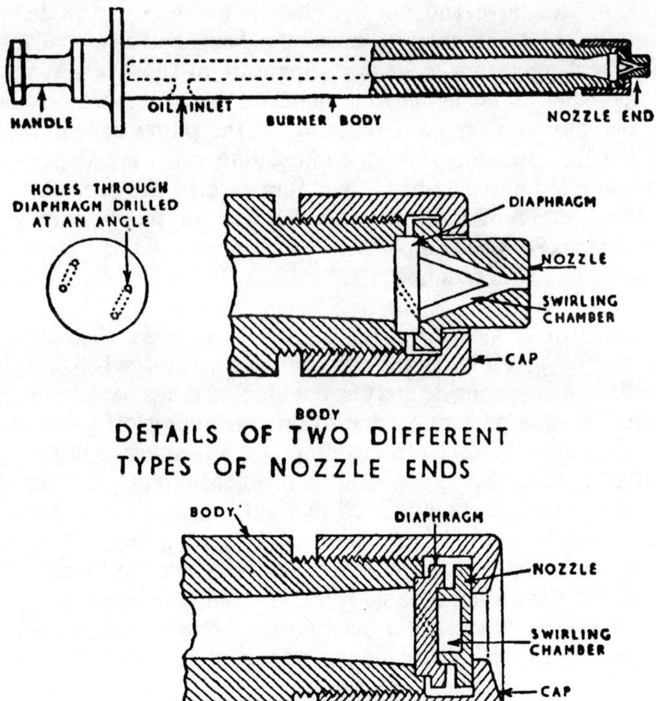

Figure 20
Wallsend-Howden oil fuel burner

furnace front and protrude into the furnace. The burner is a hollow tube with a diaphragm and nozzle in the end, the diaphragm has small holes drilled through at an angle to the axis so that the oil is broken up into a film of fine particles and acquires a spiral motion to mix readily with the air supplied over the burner as it enters the furnace.

An emergency shut off valve, usually of the quick closing type, is fitted between the settling tank and the cold filters, this valve has either an extended spindle or other remote control to the deck to enable the oil to be shut off in the case of an emergency such as a fire in the stokehold.

To light the fires, open the settling tank shut off valve, inlet and outlet valves of the cold filter, suction and delivery valves of the fuel pressure pump, inlet and outlet valves of the hot filter and boiler master valve, and ease back the circulating valve on the boiler front. Set the pump away by opening exhaust and steam valves and open heating steam to heater. Open air to furnaces. When the thermometer on the boiler front indicates the correct temperature of the oil, have an assistant standing by the pump to maintain the desired fuel pressure while the circulating valve is closed, lighted torch inserted into the furnace, and furnace oil valve opened.

The fire should light immediately. The air supply may require adjustment. For good combustion, the flame should be a short bright dazzling one with no smoke.

PRECAUTIONS BEFORE LIGHTING UP. Always make absolutely sure that there is no oil lying in the bottom of the furnace before lighting up. If there is, get inside and clean it up. Accidents have occurred by neglect of this precaution due to this oil giving off gases which, when mixed with air and heated, produce an explosive mixture, and a naked light applied could result in a violent explosion. In any case, air should always be admitted to the furnace for a few minutes before applying the lighted torch and opening the fuel valve, to sweep away the least trace of any gas there may be lodging in the furnace. If the oil from the sprayer does not ignite immediately, shut the oil valve and clean out the spilled oil from the furnace before ever attempting to light up the second time.

CAUSES OF BAD COMBUSTION. *Dirty burners*. Particles of dirt or other solid matter in the oil will partially or completely choke the fine orifices of the burners. If partially choked, oil will drip from the burner end, if fully choked the flame will extinguish. Oil supply to the faulty burner must be shut off and a spare clean burner fitted. Particular care should be taken to wipe out any loose oil in the furnace and admit air to clear out gases before relighting burner. A check on the strainers is advisable as the most likely cause is solid matter passing through faulty strainers.

Water in the oil. Sea water may find its way into the double bottom fuel tanks through hull leakage or otherwise. A small percentage of sea water in the oil will mix to form an emulsion and continue to burn. Spluttering at the burners will take place accompanied by a hissing sound, and white smoke will appear at the funnel. Firebrick-work is likely to be damaged and a froth coating form on the heating surfaces. More water in the oil will of course put the fires out. When water is first detected it is necessary to change over to another oil tank free of water.

Incorrect air pressure. An insufficient air supply will cause incomplete combustion and result in a long yellow dirty flame from the burner accompanied by black smoke. This could be due to the air pressure being too low or the air valve being partially closed. Oil to the burners must be shut off and air blown through the furnaces to remove any accumulated combustible gases before correct adjustments are made prior to relighting the burners.

An excessive air supply unnecessarily carries heat away with it to waste. It is usually indicated by white smoke at the funnel.

Incorrect oil temperature or pressure. Too low an oil temperature or too low pressure will cause dripping at the burner end and imperfect combustion. Black smoke will show at the funnel. An excessive oil temperature or excessive pressure will create unsteady burning and may develop into a panting or pulsating effect.

MAIN CAUSES OF CORROSION IN BOILERS

WATER SIDE OF BOILER.

1. *Alkalinity too low.* Feed water may be unevaporated fresh water taken on board for domestic and boiler purposes or evaporated salt water. These feed waters may be acidic causing reduction in boiler water alkalinity reserve and possible corrosion of tubes and plates.

The boiler water should be regularly tested for alkalinity reserve which must be kept within prescribed limits. If the reserve is too high some water must be blown out of the boiler and replaced with low density feed. If too low, an alkaline chemical, *e.g.* caustic soda may be added.

2. *Sea water*. This can enter the boiler feed either from a leaking condenser or priming of an evaporator. Sea water contains salts, some of which are acidic and will cause corrosion, others can lead to scale formation on heating surfaces which may then overheat and fail.

To detect the presence of sea water in the boiler a chloride test is regularly carried out.

3. *Air*. Contains oxygen which is necessary for the corrosion cycle. Air enters the feed water in open storage tanks. It may enter the system wherever pressure is below atmospheric and there is a leakage, *e.g.* condensers.

The dissolved oxygen content in the boiler water is kept to a minimum by chemical and mechanical means. Testing for dissolved oxygen entails the use of chemicals and a comparator.

GAS SIDE OF THE BOILER. *Combustion gases*. Sulphur in the fuel will burn to form sulphur oxides which in part will combine with water vapour present in the gases to give sulphuric acid vapours. If the dewpoint temperature of these vapours is reached they will condense out upon metal surfaces causing corrosion.

When steaming a boiler at reduced load, *e.g.* in port, uptake gas temperatures are low hence condensation of these acid vapours is more likely to occur. It should be noted that the dew point of sulphurous acid vapour is roughly that of the water vapour in the mixture of gases and vapours, whereas the dew point of the sulphuric acid vapour is approximately 150°C hence it may condense out upon metal surfaces at or below this temperature.

BOILER WATER TESTS. A sample of boiler water would be taken through the test cock provided (sometimes the cock is referred to as the "salinometer cock") on the boiler. The sample from water tube boilers usually passes through a cooler which, because there is no flashing off to steam, enables a more representative sample to be obtained. Flashing off to steam occurs when water at boiler pressure and temperature is suddenly released into the atmosphere.

SIMPLE WATER TESTS ARE:

1. *Density*

   A hydrometer (or salinometer) is placed in the water sample and the total dissolved solids, usually expressed in parts per million (ppm) of the sample is obtained. Density should be kept low. The higher the density the greater the possibility of scale formation on boiler heating surfaces or priming the boiler.

2. *Alkalinity*
   A strip of litmus paper is immersed into the boiler water sample. If the litmus paper changes colour to red the water sample is acidic. If it becomes a deep blue colour the water sample is alkaline.

The foregoing tests may be sufficient for some low rated domestic-auxiliary boiler but in the case of water tube and some tank type boilers more detailed, accurate analysis of boiler water condition is required. Sophisticated tests require the operator to carefully follow the instructions contained in the test kit. These tests may be used to determine: alkalinity reserve, chloride content, presence of oxygen, phosphate reserve, sulphites, hydrazine and other compounds.

OIL IN THE WATER SPACE OF A BOILER. *Sources:*

1. Oil lubricated steam machinery. Oil can enter the exhaust steam system and eventually find its way into the boiler via the condenser and feed system.
2. Steam heated fuel oil systems, *e.g.* fuel oil heaters or steam heating coils in tanks. If the tubes in the oil heater or the steam coils in the tanks leak, fuel oil can via the steam heating returns enter the condensate system and thence into the boiler.

Detection and effect. Oil would be seen in the boiler glass water gauge. As oil adheres to the heating surfaces of boiler tubes, etc, heat transfer will be affected. It will be found necessary to increase the firing rate to maintain boiler pressure and this could cause overheating and possible failure.

CHAPTER FOUR

TURBINES

Steam turbine plant consists essentially of one or two highly rated water tube boilers which supply steam to the turbines which are geared down to the propulsion shaft.

A typical plant could be two Foster Wheeler ESD boilers (ESD; – External Superheater. D shaped) supplying steam at 63.5 bar and 510°C to a single cylinder 15,000 kw impulse turbine fitted with double reduction gearing. Steam, after use in the turbine, is condensed in a regenerative condenser which is an integral part of the turbine casing, it then is pumped as feed water back to the boiler via various heaters to economise on fuel consumption and increase plant efficiency.

Some of the ancillaries associated with the plant would be, turbine or electrically driven feed pumps, centrifugal pump for condenser circulating water, closed circuit lubrication system consisting of pumps, coolers (heat is absorbed by the lubricating oil in bearings and gearing) tanks and filters.

STEAM TURBINES

A turbine, like the reciprocating engine, is a machine for converting heat energy into mechanical energy, but the principles upon which these two engines work are entirely different. In the internal combustion reciprocating engine the gas pressure acts as a static load on the piston to cause it to move and this motion is converted from a reciprocating one into a rotary motion at the shaft by connecting rod and crank mechanism. In the turbine the rotor coupled to the shaft receives its rotary motion direct from the action of high velocity steam impinging on blades fitted into grooves around the periphery of the rotor, thus the action of the steam in a turbine is "dynamic" instead of static as the gas is in the reciprocating engine.

There are two types of turbine, the *impulse* and the *reaction.* In both cases the steam is allowed to expand from a high pressure to a lower pressure so that the steam acquires a high velocity at the expense of pressure, and this high velocity steam is directed on to curved section blades which absorb some of its velocity; the difference is in the methods of expanding the steam.

In impulse turbines the steam is expanded in nozzles in which the high velocity of the steam is attained before it enters the blades on the turbine rotor, the pressure drop and consequent increase in velocity therefore takes place in these nozzles. As the steam passes over the rotor blades it loses velocity but there is no fall in pressure.

In reaction turbines, expansion of the steam takes place as it passes through the moving blades on the rotor as well as through the guide blades fixed to the casing.

Increase of pressure and temperature in modern practice has meant a move towards impulse turbines. The efficiency of the reaction turbine is reduced as steam leakage past blade tips is higher at higher pressures. The impulse reaction turbine however may still be used for low pressure turbines but the impulse turbine, being more compact and weight saving, is taking precedence.

THE IMPULSE TURBINE

As stated above, in the impulse turbine the high pressure steam passes into nozzles wherein it expands from a high pressure to a lower pressure and thus the pressure energy in the steam is converted into velocity energy (kinetic energy). The high velocity steam is directed on to blades fitted around the turbine wheel, the blades are of curved section so that the direction of the steam is changed thereby imparting a force to the blades to push the wheel around. The simplest form of impulse turbine is the single-stage De Laval shown in Fig. 21. This consists of a solid wheel bolted by flanges to a shaft, blades of bronze or nickel steel are fixed into a groove around the rim of the wheel, caulked in, and a shroud or strap wrapped around the tips of the blades to strengthen them.

The best efficiency is obtained when the linear speed of the blades is half of the velocity of the steam entering the blades, thus, when one set of nozzles is used to expand the steam from its high supply pressure right down to the final low pressure, the resultant velocity of the steam from the nozzles is very high, and to obtain a high efficiency it means therefore that the wheel should run at a very high rotational speed. Lower speeds, which are more suitable, can be obtained by *pressure-compounding*, or *velocity-compounding*, or a combination of these termed *pressure-velocity-compounding*.

In the pressure-compounded impulse turbine, the drop in pressure is carried out in stages, each stage consisting of one set of nozzles and one bladed turbine wheel, the series of wheels being keyed to

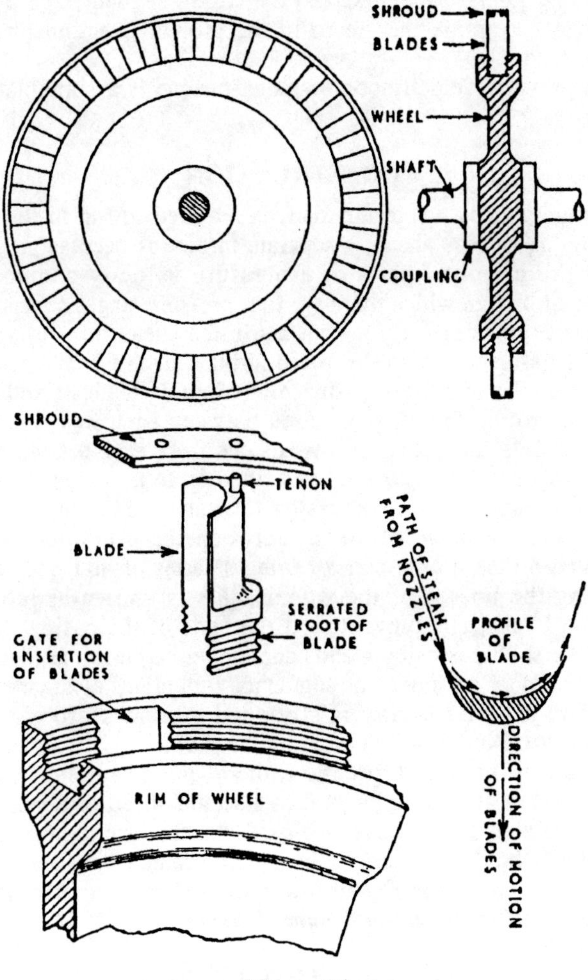

Figure 21
Single stage impulse turbine wheel

the one shaft with nozzle plates fixed to the casing between the wheels.

In the velocity-compounded impulse turbine, the complete drop in steam pressure takes place in one set of nozzles but the drop in velocity of the steam is carried out in stages, by absorbing only a part of the steam velocity in each row of blades on separate wheels and having guide blades fixed to the casing at each stage between the wheels to guide the steam in the proper direction on to the moving blades.

The pressure-velocity-compounded turbine is a combination of the two.

CONSTRUCTION

In marine engines where the shaft is required to run in the astern direction as well as ahead, a separate turbine is necessary. Fig. 22 shows the principal parts of a pressure-velocity-compounded impulse turbine in which there are four pressure stages consisting of four sets of nozzles and four wheels in the ahead turbine, and two similar pressure stages in the astern turbine. Each wheel carries two rows of blades and there is one row of guide blades fixed to the casing protruding radially inwards between each row of moving blades to drop the velocity in two steps from each set of nozzles. The wheels are of forged steel and fitted on to a mild steel stepped shaft. The nozzle plates and casing to which they are fixed, are in two halves. Leakage of steam between the pressure stages is prevented by the nozzle plates having a series of thin rings almost touching the bosses of the wheels, this is known as labyrinth packing. Leakage is prevented at the ends of the casing through which the shaft passes by glands containing carbon rings, each ring is composed of a number of segments with slight clearance at the butts and a garter spring is stretched around a groove in the periphery so that the carbon bears on the shaft.

To run in the ahead direction, the ahead steam stop valve is opened which allows steam to pass into the ahead turbine. To run astern, the ahead stop valve is closed and the astern stop valve is opened which admits steam to the astern turbine. The astern power, which is required mainly to brake the headway of the ship, is usually about 70 per cent of the ahead power.

THE REACTION TURBINE

In the type usually known as the reaction turbine, the steam is

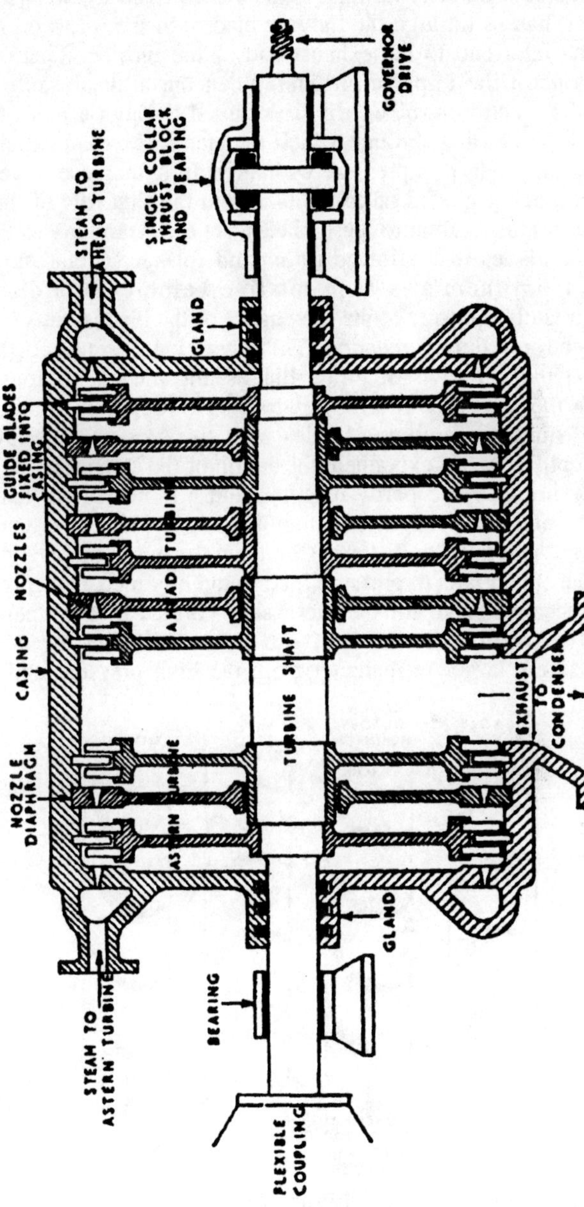

Pressure-velocity compounded impulse turbine

Figure 22

expanded continuously through guide blades fixed to the casing and also as it passes through the moving blades on the rotor, on its way from the inlet end to the exhaust end of the turbine. There are no nozzles as in the impulse turbine. When the high pressure steam enters the reaction turbine, it is first passed through a row of guide blades in the casing through which the steam is expanded slightly, causing a little drop in pressure with a resulting increase in velocity, the steam being guided on to the blades in the first row of the rotor gives an impulse effect to these blades. As the steam passes through the rotor blades it is allowed to expand further so that the steam issues from them at a high relative velocity in a direction approximately opposite to the movement of the blades, thus exerting a further force due to reaction. This operation is repeated through the next pair of rows of guide blades and moving blades, then through the next (see Fig. 23) and so on throughout a number of rows of guide and moving blades until the pressure has fallen to exhaust pressure. As explained, the action of the steam on the blades is partly impulse and partly reaction, and a more correct name for this type of turbine might be "impulse-reaction" but it is generally known as "reaction" to distinguish it from the pure impulse type.

As the steam falls in pressure it consequently increases in volume and to accommodate for the increasing volume of the steam as it passes through the turbine, the rotor and casing is made progressively larger in diameter from the high pressure end to the

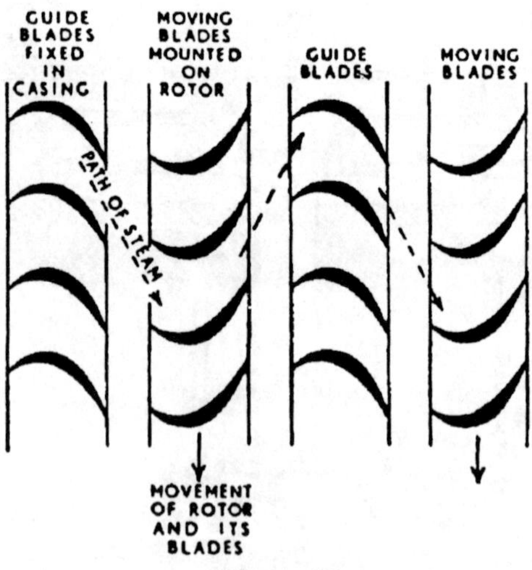

Figure 23
Principle of reaction blading

low pressure, usually in steps, with larger area of annulus and longer blades.

ASTERN RUNNING

As in all turbine installations on board ship, a separate astern turbine is required to drive the propeller in the reverse direction. This astern turbine may be mounted on the same shaft as the ahead turbine, it may be a completely separate unit geared to the main shaft, or the lay-out may be h.p. and i.p. ahead turbines on one turbine shaft and another shaft in parallel carrying the l.p. ahead and the astern turbines, each shaft being geared to the main shafting (see Figs. 22 and 24). The ahead and astern turbines have their own separate steam stop valves, or it could be one steam chest with two valves and branches, specially designed so that it is not possible to open both ahead and astern steam valves at the same time.

GEARING

The speed at which turbines should run to obtain the best efficiency, is high, much higher than the economical speed of a ship's propeller, therefore to obtain the best performance from both turbine and propeller, reduction gearing is interposed between the turbine shaft and the propeller shaft, to enable both to run at their best speeds.

Single reduction gearing consists of one stage of speed reduction in which a pinion on the turbine shaft meshes with a gear wheel on the main shaft.

Double reduction gearing consists of two stages of speed reduction by means of a pinion on the turbine shaft meshing with an intermediate wheel, on the intermediate shaft a secondary pinion meshes with the main gear wheel on the propeller shafting.

The teeth in the wheels are cut at an angle to the axis of the shaft (these are known as helical teeth) for smooth running, and all gear wheels are arranged in pairs to balance any axial thrust caused by the teeth being helical. Fig. 24 shows diagrammatically a typical turbine lay-out including astern turbine and double-reduction gearing. Note the flexible couplings between turbine shafts and pinion shafts, these are to prevent thermal expansion of the turbine or misalignment of its shaft affecting the perfect meshing of the gear wheel teeth.

REGENERATIVE CONDENSER

The function of a condenser is to condense the exhaust steam from the engine into water. The disadvantage of an ordinary condenser is

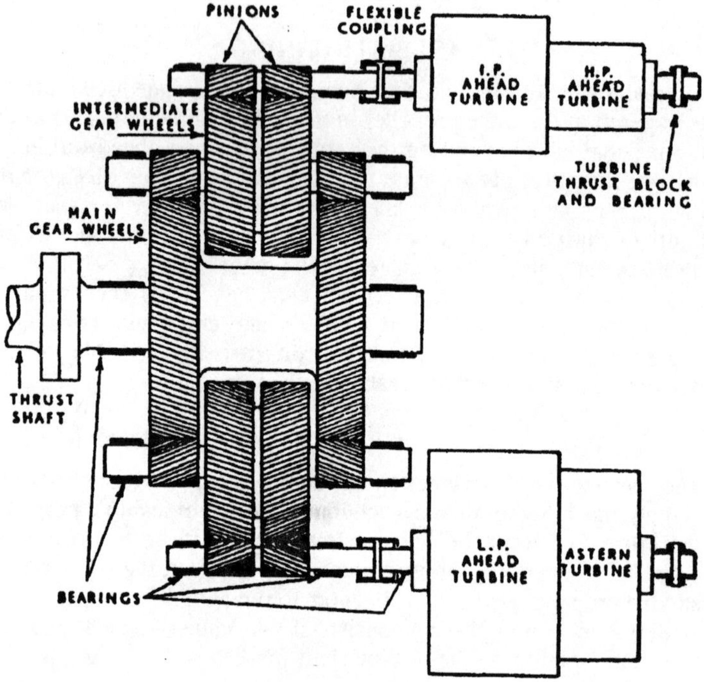

Figure 24
Double-reduction gearing

that it goes beyond this requirement and cools the water to a lower temperature after it has been condensed, this is known as overcooling of the condensate and constitutes a loss of efficiency as the lost heat must be made up again at the expense of fuel. The regenerative condenser is designed so that the condensate leaves the condenser as near as possible to the temperature of the exhaust steam entering, and therefore in effect, only the latent heat is extracted from the steam.

Fig. 25 shows a section through a regenerative condenser. As the steam condenses, it drops on to sloping baffles and falls off like rain into a well at the bottom of the condenser. A gap is left between the nests of tubes to allow some of the exhaust steam to pass directly to the bottom part of the condenser where it mixes with the water droppings and surface of the water in the well, this heats up any water which is below the steam temperature; as this free passage steam condenses when it gives up its heat to the water, it falls itself as water to the bottom, any free steam not so condensed tends to rise towards the air-extraction outlet and condenses immediately on coming into contact with the lower tubes. A condensate pump extracts the condensed water, the air is extracted separately by a steam-jet air extractor, thus a very high vacuum is maintained which

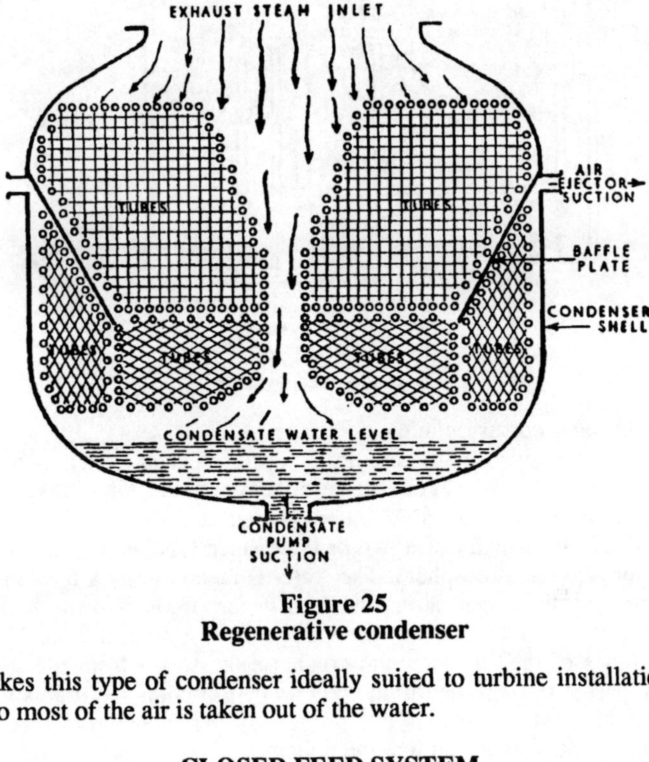

Figure 25
Regenerative condenser

makes this type of condenser ideally suited to turbine installations; also most of the air is taken out of the water.

CLOSED FEED SYSTEM

In modern steam turbine installations with water-tube boilers, the feed water circuit from the regenerative condenser to the boilers is completely closed. This is known as the *closed feed system* and its essential function is to prevent the condensate, from which most of the air has been extracted in the condenser, from coming into contact with the atmospheric air before being returned as feed water to the boilers. The de-aerated water would absorb air if any part of the circuit was open to the atmosphere and the oxygen of the air

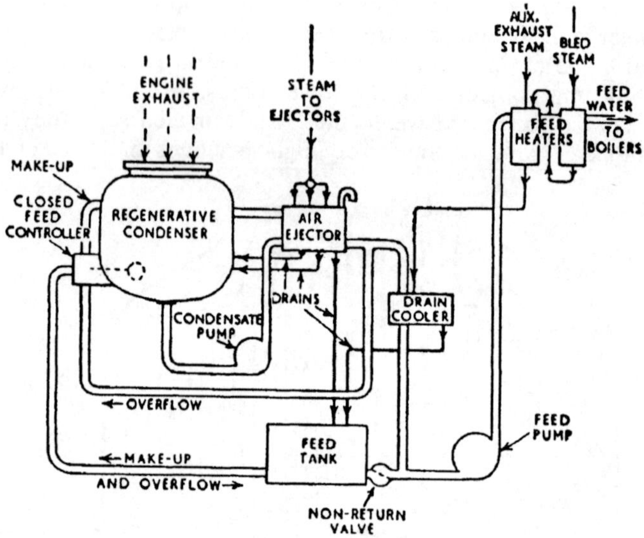

Figure 26
Closed feed system

would cause corrosion in the boilers.

Fig. 26 shows diagrammatically a closed feed system with its basic units. In the regenerative condenser, the air is taken out separately to the condensed water. The air is extracted by a steam-jet air ejector usually of a two or three stage type which discharges the air into the atmosphere. The water is taken out by a condensate pump and discharged along the feed water circuit into the boilers. Along the feed water path, the water passes around the outside of the tubes of the air ejector to condense the steam from the steam jets, passes through the drain cooler where it cools the drains from the feed heaters, then is discharged by the feed pump through surface feed heaters and into the boilers.

Drains and make-up feed water are led to the condenser where the water is de-aerated before passing into the feed range. The level of the condensate in the bottom of the condenser is kept constant by a float operated control valve (closed feed controller) which allows make-up water to be drawn into the condenser from the feed tank, or to be discharged into the feed tank from the condenser.

WARMING THROUGH PRIOR TO STARTING

The turbines must be thoroughly heated up as near as possible to working temperature conditions before starting. The practice varies considerably depending upon the type of installation, size and design of turbines, one common procedure however is as follows, assuming the auxiliary steam line is already in use:

Ascertain that all drains on the engines and on the main steam line are open.

Start the lubrication pumps, check the oil pressures at the bearings and see that the gearing sprayers are all clear and working properly.

Open main injection and discharge valves and start main condenser circulating pump on slow running.

Run the condenser extraction pump to maintain a low vacuum of about 200mm Hg.

Change over auxiliary exhausts from auxiliary condenser to main condenser.

See that propeller is free to be turned without obstruction by ropes or small craft. Turn the engines a little then take out the turning gear.

Ease main steam valve off its face to allow steam into main steam pipe. After waiting until any water in the line is driven out through the drains, open the valve *very slowly* and when the drains are blowing freely they are (gradually) closed.

Check that there is no obstruction to sliding feet and take expansion readings.

Put on gland steam and crack open manœuvring valves, easing sufficiently to allow a little flow of steam through the engines.

After about 2½ hours raise vacuum to about 350mm Hg and operate manœuvring valves carefully to slowly turn the engines a few revolutions ahead and then astern, repeating this operation about every 10 or 15 minutes for the next 1½ hours.

At ready for Stand By, after at least four hours of heating up, pumps are brought up to normal running speed, vacuum lifted up, and expansion readings checked.

During manœuvring, the turbine drains are regulated as required, and shut off at full-away.

TURBINE PLANT ALARM AND SAFETY DEVICES.

Protection of expensive high pressure steam turbine plant and that of the personnel is important. In modern plant automatic shut down can occur which in some cases can be problematical viz; loss of motive power when manœuvring.

Some of the safety devices and alarms are: –

1. Boiler safety valves. Which relieve overpressure.
2. High and low boiler water level alarms.
3. Emergency self closing stop valve.
   This valve is arranged to close automatically and shut off steam supply to the turbine as a result of.
   a. Loss of lubricating oil pressure.
   b. Turbine overspeed.
   c. Excessive axial rotor travel. (if not prevented, blade damage could occur).
   d. Low condenser vacuum.
4. Condenser circulating water alarm.
5. Turning gear in alarm.
6. Indication of electrical failure.

The reader should refer to Chapter 11 for further information regarding the monitoring of turbine machinery.

MEASUREMENT OF POWER

The power output of a turbine is measured at the main shafting. A special device is fitted on the shaft known as a *torsionmeter* to measure the angle of twist in the shaft over a given length. The shaft is always slightly twisted when transmitting power and the angle of twist is proportional to the power transmitted. This information together with the rotational speed of the shaft is multiplied by a constant to obtain the shaft power.

Consider a shaft being turned by a steady force of F newtons applied at a leverage or radius of L metres.

Work done to turn shaft one revolution

$$= \text{force [N]} \times \text{distance [m]}$$
$$= F \times 2\pi L \text{ [newton-metres = joules]}$$

If the shaft is running at n revolutions per second, work done per second = power

$$= F \times 2\pi L \times n \text{ [joules per second = watts]}$$

but,

F[N] x L[m] $= \text{Torque [N m]}$

$\therefore$ Power $= 2\pi Tn$ [W].

The torque T in the shaft cannot be conveniently measured directly while it is running but measurement of the angle of twist is comparatively simple.

Since the angle of twist is proportional to the torque, a test on the shaft is performed before installation to obtain the relation between torque and angle of twist over a chosen length. The shaft is gripped and held firmly at one end and a torque (twisting moment) applied at the free end to ascertain the torque required to twist the shaft over the given length through one degree.

If t = torque to produce 1° angle of twist,
$t\theta$ = torque to produce $\theta°$ angle of twist.

Hence when the shaft is running and it is twisted θ degrees, the torque in the shaft is $t\theta$ and this can replace T in the above expression:

$$\text{Shaft power} = 2\pi t\theta n$$

In this, 2π and t are constants, therefore let C represent $2\pi t$ thus:

$$\text{Shaft power} = C\theta n$$

So that when the ship is at sea it only remains to read the angle of twist as recorded by the torsionmeter, note the speed of the shaft in revolutions per second, then multiply the product of these by the constant C to obtain the shaft power.

As previously stated, the angle of twist is measured by a torsionmeter. There are various designs in use, the most common being the electrical type.

GAS TURBINES

Marine type gas turbines work on the constant pressure (joule) cycle. An open-cycle gas turbine plant consists of three essential parts – *air compressor*, *combustion chamber*, and *turbine*. Fig. 27 is a diagrammatic sketch of a simple plant which includes a *heat exchanger*. Air is drawn into the compressor from the atmosphere and is compressed to a higher pressure (pressure ratio about 5 or 6 to 1), smaller volume and higher temperature. This compressed air passes through the heat exchanger where some of the heat energy in the exhaust gases (which would otherwise go to waste) is transferred to the air and further increases its temperature and volume. The air now enters the combustion chamber, in here some of the air is admitted through the burner into the combustion chamber and is used for burning the fuel, the remainder of the air passes through the jacket surrounding the burner housing, mixes with the products of combustion, and is heated at constant pressure, the temperature and volume thereby increasing. The mixture of hot air and gases now passes through the turbine where it does work in driving the rotor as it expands to almost atmospheric pressure. Finally the gases exhaust via the heat exchanger to the atmosphere.

Of the power developed in the turbine, some is absorbed in driving the compressor, the remainder being available for external use such as for propulsion or driving an electric generator. A starting motor is fitted at the opposite end of the shaft to that for the external drive.

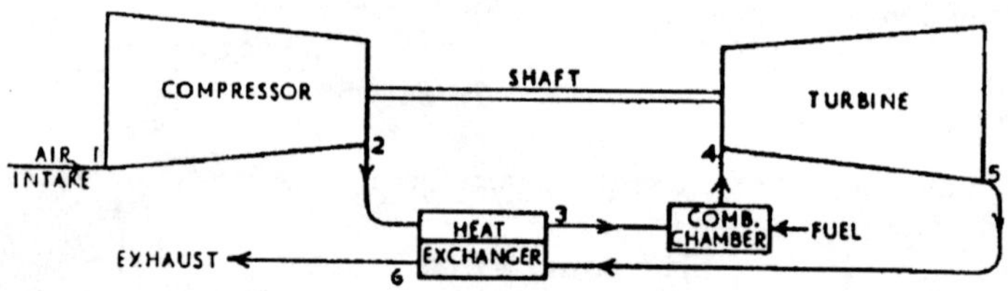

Figure 27

CHAPTER FIVE

INTERNAL COMBUSTION ENGINES

Engines falling into this category are so named because combustion of the fuel takes place *inside* the engine, hence *internal* combustion, as compared with steam engines where the fuel is burned outside the engines, *i.e.*, in the boilers. When the fuel burns inside the engine cylinder, it gives out heat which is absorbed by the air previously taken into the cylinder, the air consequently expands and increases in pressure and imparts its energy to the piston. The method of igniting the fuel varies. In diesel engines the air in the cylinder is compressed to a high pressure so that it attains a high temperature, the oil fuel is injected into this high temperature air and the fuel immediately ignites. When the ignition of the fuel is caused by this method the engine is classed as a *compression-ignition* engine. In petrol engines the fuel is most often ignited by an electric spark.

In the sketches to follow illustrating the working cycles of the four-stroke and two-stroke diesel engines, the usual simplified diagrammatic form is used by not including a piston rod, that is, by showing the connecting rod directly connected between piston and crank pin as it is in small high speed engines.

The reciprocating motion of the piston is converted into a rotary motion at the crank shaft by connecting rod and crank. As stated above, in small high speed engines such as for motor cars and lifeboats which are always single-acting, there is no piston rod. The connecting rod has an eye at its top end and a gudgeon pin passes through this and the sides of the hollow piston to form a swivel connection between rod and piston, and the bottom of the cylinder is open to allow for the swing of the connecting rod. The piston therefore performs the additional function of a crosshead and the side thrust at the top of the connecting rod is taken up on the cylinder wall.

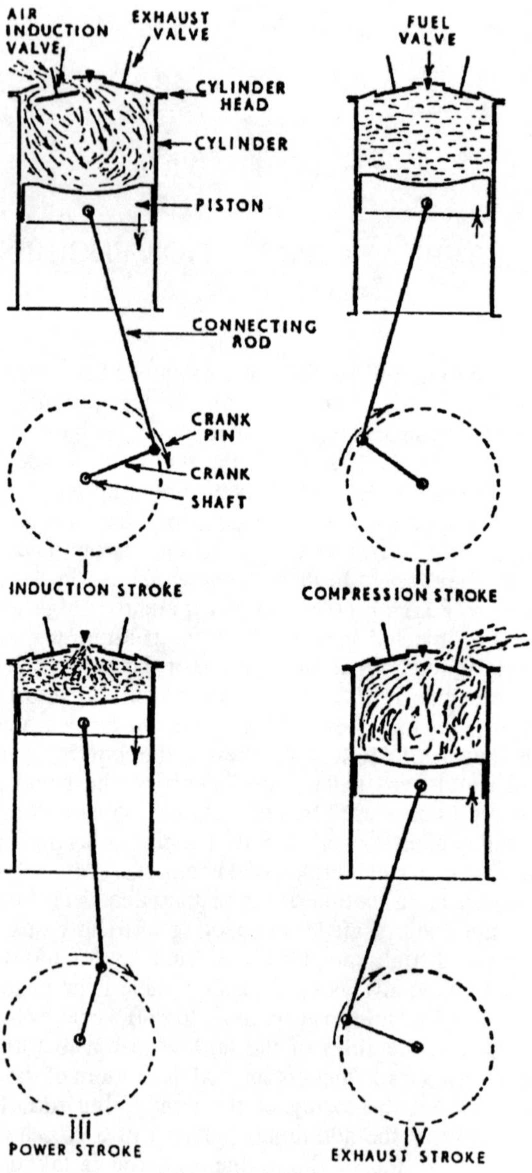

Figure 28
Four-stroke diesel engine

In large engines a piston rod and crosshead are fitted so that the side thrust is taken up by the crosshead guide shoes on the guides.

THE FOUR-STROKE DIESEL ENGINE

This engine is so named because it takes four strokes of the piston (that is two revolutions of the crank shaft) to complete one working cycle of operations.

Fig. 28 illustrates each of these four strokes in one cylinder. On the cylinder head is shown the fuel valve which lifts to admit oil fuel into the cylinder, the air-induction valve through which air is drawn in, and the exhaust valve through which the exhaust gases are expelled from the cylinder. There are two more valves which are not shown here because they do not operate during the normal working cycle, one is the relief valve, spring loaded like a safety valve, which will open when the pressure in the cylinder rises too high, the other is the air starting valve which opens to admit compressed air to start the engine.

CYCLE OF OPERATIONS IN A FOUR-STROKE DIESEL ENGINE

Sketch (i). This illustrates the INDUCTION STROKE. The piston is moving down, the air induction valve is open and air is being drawn into the cylinder from the atmosphere by the suction effect of the piston. At about the end of this stroke the cylinder is full of air and the air induction valve closes.

Sketch (ii). This shows the COMPRESSION STROKE. The piston is moving up, no valves are open and therefore the air in the cylinder is being compressed. When air is compressed its temperature rises and the reason for compressing the air in a diesel engine is to obtain a sufficiently high temperature to cause the oil to ignite and burn rapidly when it is injected into the cylinder at the end of this stroke. The pressure of the air at the end of compression is in the region of 35 bar to obtain a temperature of about 540°C.

Sketch (iii). This is the POWER STROKE and the piston is moving down. The fuel is injected into the cylinder in the form of a fine spray through the fuel valve, it mixes with the hot air and burns rapidly; the fuel is admitted a few degrees before top dead centre of the crank to give it time to reach full combustion for the beginning of the stroke and the valve remains open for about one-tenth of the downward stroke. As the oil burns it heats the air which causes it to rise in pressure or increase in volume; in the pure diesel engine the admission of the oil is controlled so that the pressure of the gases remains constant while the piston is moving down during the combustion period (for this reason it is called the constant-pressure cycle), in actual practice the pressure rises a little during combustion. When the fuel is shut off, the gases continue to push the piston forward and the pressure consequently falls towards the end of the stroke, this is the expansion period of the power stroke. At the end of fuel combustion (and beginning of expansion) the temperature of the gases has probably risen to about 1,650°C. Near the end of this stroke, when the pressure has fallen to be of little further use, the exhaust valve opens.

Sketch (iv). This illustrates the EXHAUST STROKE, the exhaust valve is open, piston moving up, and the gases are being expelled from the cylinder. At the end of this stroke the exhaust valve closes and the air-induction valve opens to begin the cycle of operations over again.

There is a certain amount of overlapping with regard to the closing of the exhaust and opening of air-induction, for a short period they are both open together; the momentum of the exhaust gases sweeping out through the exhaust valve has the effect of pulling in air through the induction valve and so assists in scavenging the combustion space at the beginning of the air-induction stroke.

The timing diagram shown in Fig. 29 is one example of the timing of the opening and closing of the valves with respect to the position of the crank, in a four-stroke diesel engine, these figures vary in different engines.

Fig. 30 is an exaggerated indicator diagram which shows the variation of pressure against volume throughout the cycle.

VALVE MECHANISM

The air-induction, exhaust and, in some cases, starting-air valves are opened by means of rocking levers fulcrummed about their centre and actuated by cams fixed to the cam shaft. Each cam has a peak, which, when it comes around to contact the roller on the end of the rocking lever, pushes this end up, and the other end is depressed to open the valve against a spring in the valve housing (see Fig. 31). Push rods may be used as distance pieces between cams and rocking levers. The cams are set in the correct position relative to the crank so that the valves open and close at the exact moment and for the required period in the working cycle. As the timing of opening and closing of the valves is relative to the crank position and direction of

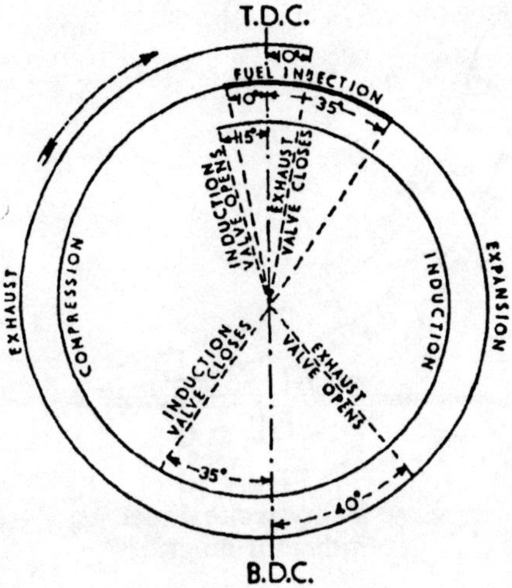

Figure 29
Timing diagram for a four-stroke diesel engine

movement, separate cams fixed at different relative positions are required to actuate the rocking levers to run the engine in the reverse direction.

The working cycle constitutes four strokes of the piston, which is two revolutions of the crank shaft. During one cycle each valve is opened only once, hence in the four-stroke engine the cam shaft is driven at half the speed of the crank shaft.

The fuel valve is usually opened by the pressure of the oil discharged by the fuel pump and closes under the action of a spring when the pressure is released. The timing of the beginning of opening, period of opening, and closing of the valve is varied by the fuel pump plunger.

THE TWO-STROKE DIESEL ENGINE

The two-stroke diesel engine is so named because it takes two strokes of the piston to complete one working cycle. Every downward stroke of the piston is a power stroke, every upward stroke is a compression stroke, the exhaust of the burned gases from the cylinder and the fresh charge of air is taken in during the late

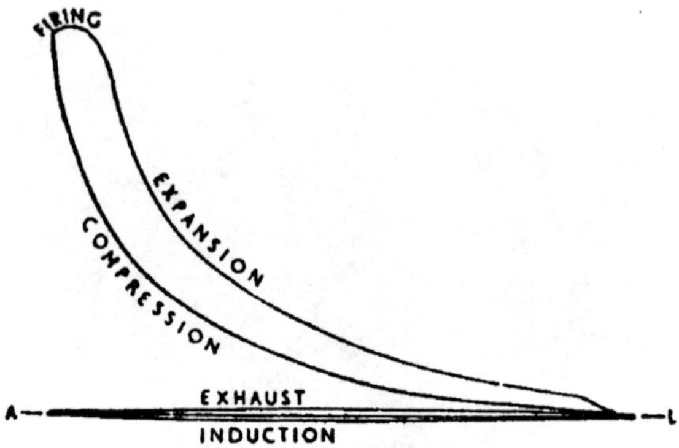

Figure 30
Four-stroke diesel indicator diagram

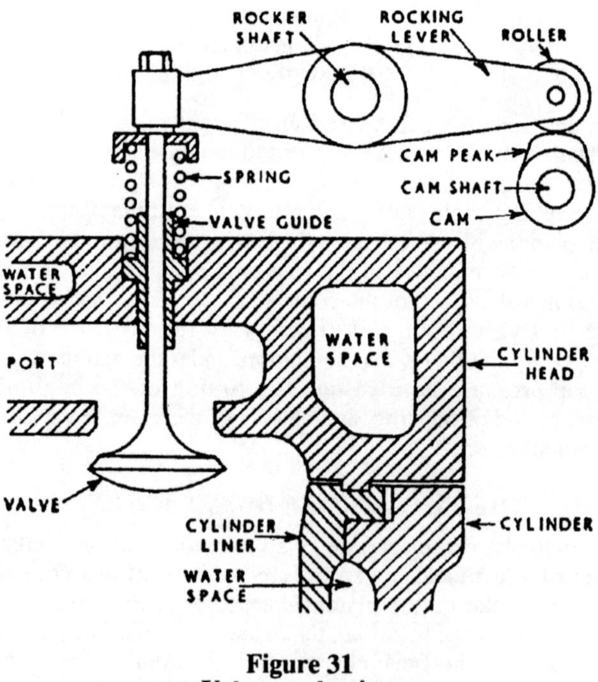

Figure 31
Valve mechanism

period of the downward stroke and the early part of the upward stroke. In the basic two-stroke engine, the exhaust gases pass through a set of ports in the lower part of the cylinder and the air is admitted through a similar set of ports, the ports are covered and uncovered by the piston itself which must be a long one or have a skirt attached so that the ports are covered when the piston is at the top of its stroke.

As there is no complete stroke to draw the air into the cylinder, the air must be pumped in at a low pressure from a pump, this is known as the scavenge pump, the air supplied is referred to as scavenge-air and the ports in the cylinder through which the air is admitted are termed scavenge ports. It is the function of this air to sweep around the cylinder and so "scavenge" or clean out the cylinder by pushing the remains of the exhaust gases out, leaving a clean charge of air to be compressed.

Normally therefore there is no air-induction valve or exhaust valve in the cylinder head as there are in a four-stroke engine and the cylinder head is consequently a much simpler and stronger casting; there are of course the fuel valve, air-starting valve and relief valve.

In some two-stroke engines, exhaust may be through ports near the top of the cylinder, opened and closed by a piston; or through a mechanically-operated valve in the cylinder cover. Both these methods give a better scavenging effect.

As the cycle of operations takes place in two-strokes of the piston, which is one revolution of the crank shaft, the cam shaft is driven at the same speed as the crank shaft.

CYCLE OF OPERATIONS IN A TWO-STROKE DIESEL ENGINE

The sketches in Fig. 32 illustrate various points in the working cycle in one cylinder.

Sketch (i) shows the piston moving up, the exhaust and scavenge ports are covered by the piston and the fuel valve is shut. Air previously taken into the cylinders is being compressed to about 35 bar and 540°C at the end of compression.

Sketch (ii) illustrates the fuel being injected into the cylinder, which, being broken up into the form of a fine spray, readily mixes with the hot air, burns and gives out heat. The fuel is injected at such a rate that the pressure of the gases inside the cylinder is kept constant (or may rise slightly in actual practice) during combustion as the piston moves forward. When the fuel is cut off at about one-

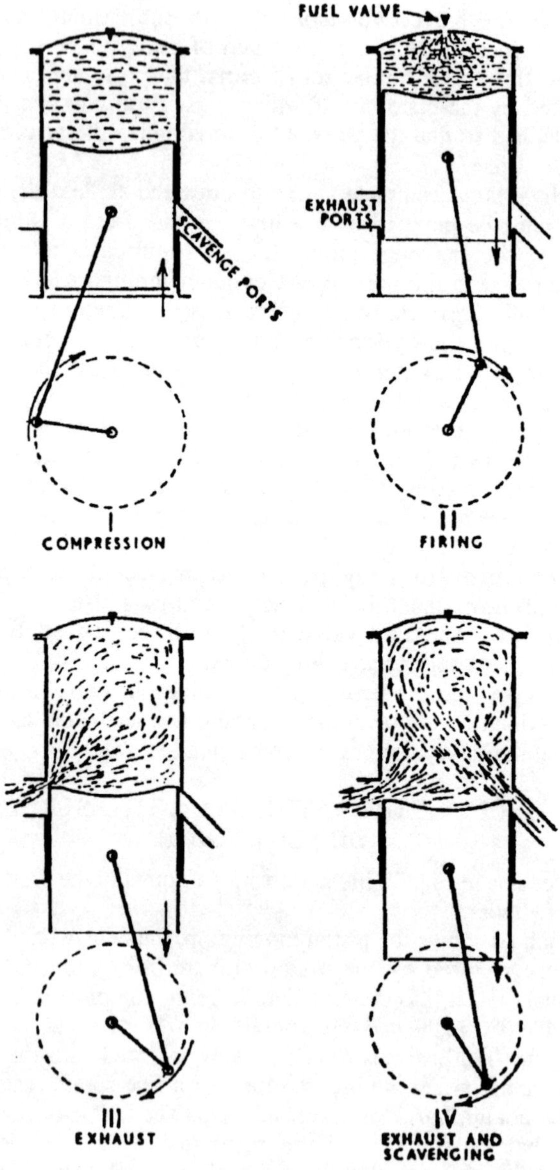

Figure 32
Two-stroke diesel engine

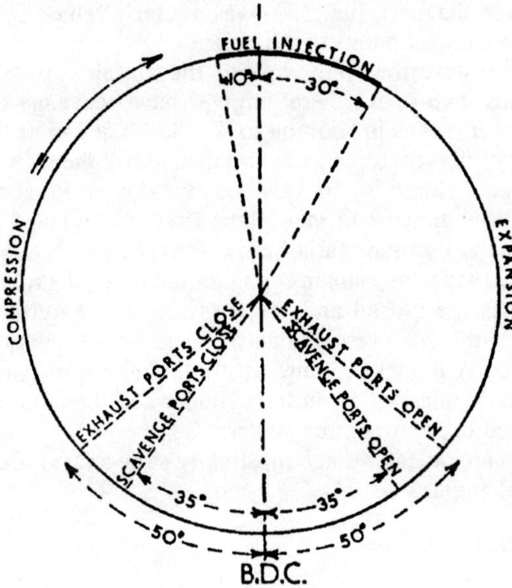

Figure 33
Timing diagram for a two-stroke diesel engine

tenth of the downward stroke, the hot gases contain sufficient energy to continue to do work on the piston and push it forward towards the end of the stroke, the gases consequently falling in pressure as they expand.

Sketch (iii) The piston is still moving down and has just begun to uncover the exhaust ports (note that the top of the exhaust ports are at a slightly higher level than the top of the scavenge ports), the first rush of exhaust gases out of the cylinder is taking place, whatever pressure there was now rapidly falls to about zero.

Sketch (iv) The piston has moved further down to uncover also the scavenge ports, the scavenge air at a little above atmospheric pressure sweeps into the cylinder. Note the slope of the air passage which directs the air upwards into the cylinder. The scavenge ports are cut through partially tangential to the cylinder to give the air a swirling movement. The cylinder is now being scavenged by expelling all the burned gases out.

When the crank passes bottom dead centre, the piston moves up and covers the scavenge and exhaust ports, the air trapped in the cylinder is then compressed to begin the cycle over again.

The timing diagram, Fig. 33, gives average values of the crank angles at the cardinal points of the cycle.

The above describes the cycle of the simplest of two-stroke engines. Most two-stroke diesel engines have scavenge control by valves or other means in addition to the sleeve action of the piston, an extra set of scavenge ports are arranged above the main scavenge ports which are closed by the valves on the downward stroke of the piston to allow exhaust to take place first, and opened when the pressure in the cylinder falls below scavenge pressure. By this means the scavenge air remains open on the upward stroke after the exhaust ports are closed and the pressure in the cylinder at the beginning of the compression can be equal to the pressure of the scavenge air. A higher pressure at the beginning of compression means a greater mass of air in the cylinder and therefore more fuel can be burned to produce more power.

Fig. 34 is an indicator diagram, slightly exaggerated, from a two-stroke diesel engine.

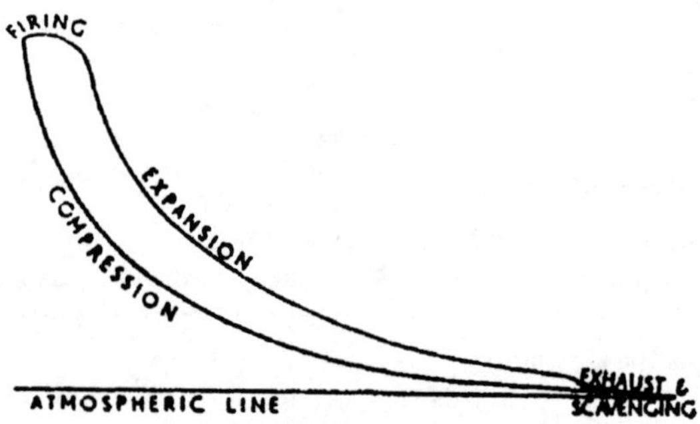

Figure 34
Two-stroke diesel indicator diagram

A diesel engine system may consist of:-

1. A single direct drive, multi-cylinder, slow speed supercharged two stroke cycle diesel engine running at about 60 to 110 revolutions per minute, or 2. Two (or more) medium speed, geared, multi-cylinder, supercharged two stroke cycle diesel engines running at about 200 to 300 revolutions per minute.

1.and 2. would as part of the layout require fuel, lubrication and cooling systems. Fig. 35 shows in simplified diagrammatic form a medium speed diesel engined arrangement.

Some advantages claimed are: –

a) Compact, low weight.
b) Gearing frees Engine designers and Naval Architects to produce best power-weight ratio engine and propeller respectively.
c) Simpler control. Engines can be unidirectional constant speed, hence no complex reversing mechanism or need for stopping and starting.Control is achieved by the controllable pitch propeller.

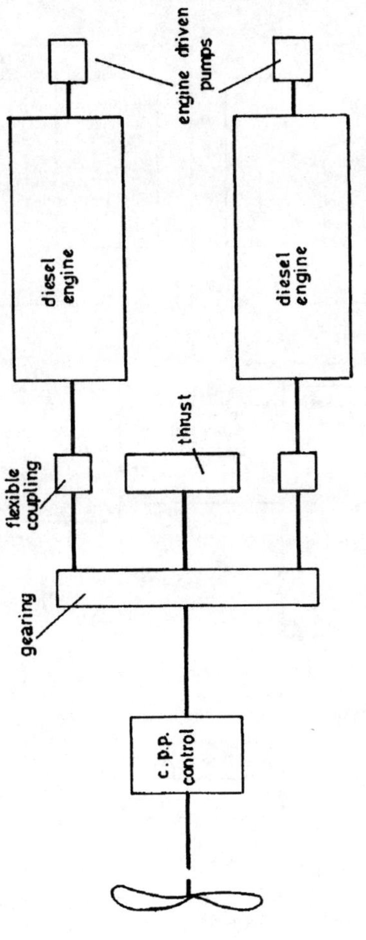

Figure 35
Medium speed diesel engined arrangement

FUEL SYSTEM

Most diesel engine fuel systems are designed to use residual oil (which requires heating) for normal running, and diesel oil for abnormal conditions and sometimes manœuvring.

Referring to Fig. 36 which shows, in a simplified form, a typical fuel system, oil is pumped up from the double-bottom tank to a settling tank. The settling tanks are in duplicate and fitted with internal heaters, so that when one tank is in use, the oil in the other tank can be settling and impurities drained off, the settling process being accelerated by moderate heating. The oil passes from the

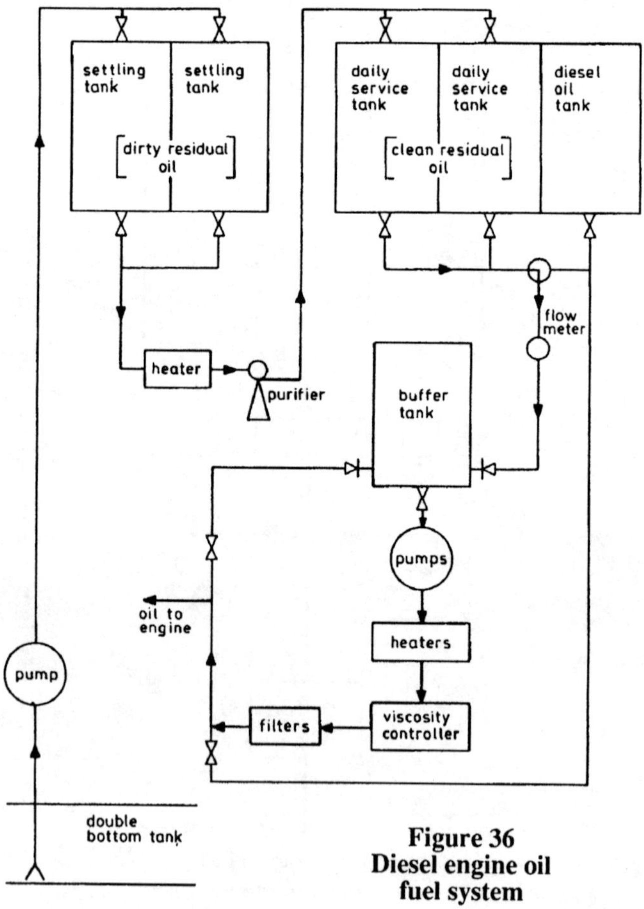

Figure 36
Diesel engine oil fuel system

settling tank in use, through a heater unit to centrifugal purifiers and then to the daily service tank. The daily service tanks are also duplicated so that one can be filling whilst the other is in service. From here the oil passes through a flow-meter and into a buffer or circulating tank. A booster pump draws the oil from this tank and discharges it through a heater, viscosity controller, filter, and finally to the engine fuel pumps. The engine fuel pumps deliver the oil into the cylinders via the fuel injectors.

SCAVENGING

We have seen that a two-stroke diesel engine must have some form of pumping unit to supply air to scavenge the cylinders of the products of combustion and recharge them with clean air. Scavenge pumps may be either reciprocating or rotary, both types usually derive their operative power from the main engine.

A reciprocating scavenge pump may be driven direct by an additional crank on the engine crankshaft (this with its additional cylinder increases the length of the engine), or it may be operated by links attached to a crosshead of the engine. The air is delivered by the pump unit at about 1.17 bar into feeder trunking which feeds each engine cylinder through its scavenge ports in turn with the necessary air for scavenging and recharging at the correct time in the cycle.

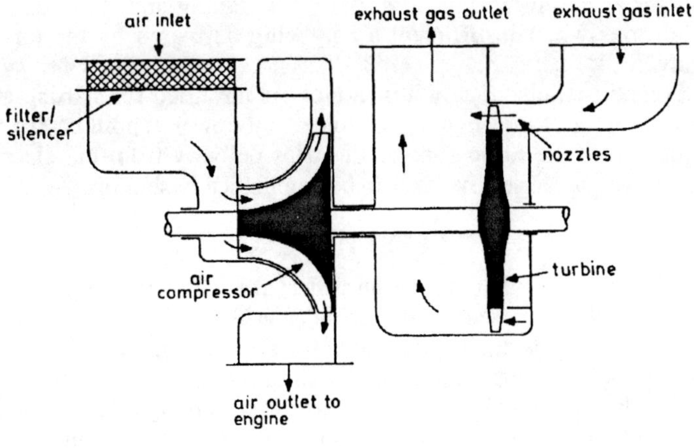

Figure 37
Exhaust gas turbo-charger

Rotary scavenge pumps (sometimes called blowers) are chain or gear driven from the engine crankshaft. These have the advantages of steady air delivery and compactness since their rotational speeds are high and constant during normal operation. They have the disadvantage in reversible engines of requiring a changeover valve since their direction of rotation is geared to that of the engine.

Another method of scavenging is by a turbo-charger (or turbo-blower). This is a turbine-compressor, the turbine uses energy from the engine exhaust gases and drives an air compressor which supplies air to the engine for scavenging and supercharging. Referring to Fig. 37 the main structure of an exhaust gas turbo-charger may consist of four separate casings which can be bolted together in various positions to suit different arrangements of engine ducting. Exhaust gases from the diesel engine enter the single stage gas turbine through the water cooled cast iron inlet casing, expand in the nozzles thereby gaining in velocity, and pass through the turbine blades whilst driving the turbine rotor. The exhaust gases leave the turbine through the water-cooled outlet casing and flow to the atmosphere, in some cases via a waste heat boiler. Combustion air for the diesel engine enters the centrifugal air compressor through a silencer-filter. The air is compressed and delivered from the compressor to an air cooler and then to the engine cylinders.

In some engines a scavenge pump unit is formed by sealing the under piston space from the crank-case and fitting appropriate valves. This unit is usually incorporated in series with a turbo-charger as a booster unit, it is useful for starting and slow running conditions when insufficient air is being delivered by the turbo-charger.

Sometimes, electrically driven blowers are fitted for starting and slow running of the engine, then under full power conditions the air supply from the turbo-charger outstrips delivery from the electric blower which causes the latter to be automatically shut off.

SUPERCHARGING

The mass of fuel that can be burned in the cylinder depends upon the mass of air present in the cylinder at the end of compression. Hence, by increasing the pressure of the scavenge air in a two-stroke engine, and by supplying air under slight pressure in a four-stroke engine during the induction stroke instead of relying on drawing the air in by suction, a greater mass of air for compression can be supplied to the engine. More fuel can then be burned without causing excessive temperature during combustion. Burning more fuel produces more heat energy to be imparted to the piston, thereby increasing the power of the engine. When an extra pressure of air is supplied so that more fuel can be burned, the engine is said to be *supercharged.*

The air supply pressure in supercharged engines varies, but is often around 200 millibars above atmospheric pressure.

SCAVENGE FIRES

Accumulation of oil and dirt can occur in the scavenge spaces of a diesel engine, this could be caused by such faults as excessive cylinder lubrication, slack, worn or broken piston rings, uneven cylinder liner wear, damaged air inlet filters, and cracked oil-cooled pistons.

If flames from combustion in the cylinder can blow past the piston into the scavenge trunking, this accumulation of oil and dirt could be ignited, thus resulting in a scavenge fire.

Indications of a scavenge fire are excessive black smoke, high exhaust temperature, paint blistering and peeling from scavenge trunking. Normally this would be local, that is, confined to the vicinity of one cylinder. The fuel should be turned off the affected cylinder unit (this reduces temperature), cylinder lubrication should be increased to minimise the risk of seizure and the engine slowed down to reduce air supply to the fire. If the engine is stopped, overheated parts may become distorted, seizure may result and the fire may spread. The fire will probably burn itself out but in the most serious cases it may be necessary to stop the engine and fight the fire. Some diesel engines have fire detection and $C0_2$ gas smothering systems for the scavenge spaces.

CRANKCASE EXPLOSIONS.

Three things are required: – fuel,air,source of ignition.

In the crankcase of a diesel engine lubricating oil provides the fuel,air is always present there so the final ingredient required is a "hot spot".

Friction between moving lubricated components in the crankcase, *e.g.* an overheated bearing, can result in a "hot spot"which vapourises lubricating oil to form an oil mist.

This oil mist with air in the correct ratio can ignite at the "hot spot" producing a primary explosion.

The pressure generated by the resultant flame speed will increase unless relieved. Guarded (to protect personnel) non-return pressure relieve valves are normally fitted to the crankcase of each cylinder and to the gear case. If after the primary explosion air enters the crankcase, due to the partial vacuum created by the rushing out of the hot gases, a secondary more violent and dangerous explosion may result with disastrous consequences. After a crankcase explosion and assuming damage has not resulted to allow ingress of air and a secondary explosion, the engine must be stopped and allowed to cool down. On no account must the crankcase doors be removed until the engine has cooled down, as the ingress of air may result in another explosion.

CRANKCASE OIL MIST DETECTOR

If condensed oil mists are the sole explosive medium then photo-electric detection should give complete protection but if the crankcase spray is explosive the mist detection will only indicate a potential source of ignition. The working of one design of detector should be fairly clear from Fig. 38 The photo cells are normally in a state of electric balance, *i.e.* measure and reference tube mist content in equilibrium. Out of balance current due to rise of crankcase mist density can be arranged to indicate on a galvanometer which can be connected to continuous chart recording and auto visual or audible alarms. The suction fan draws a large volume of slow moving oil-air vapour mixture in turn from various crankcase selection points. Oil mist near the lower critical density region has a very high optical density. Alarm is normally arranged to operate at 2½% of the lower critical point, *i.e.* assuming 50 mg/1 as lower explosive limit then warning at 1.25 mg/1.

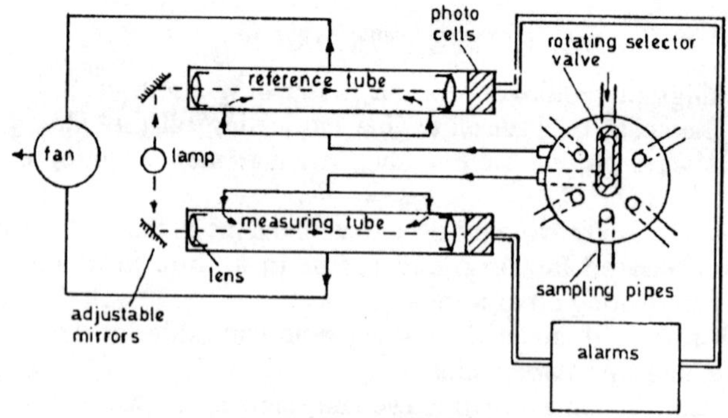

Figure 38
Crankcase oil mist detector

Operation

The fan draws a sample of oil mist through the rotary valve from each crankcase sampling pipe in turn, then through the measuring tube and delivers it to atmosphere. An average sample is drawn from the rotary valve chamber through the reference tube and delivered to atmosphere at the same time. In the event of overheating in any part of the crankcase there will be a difference in optical density in the two tubes, hence less light will fall on the photo cell in the measuring tube. The photo cell outputs will be different and when the current difference reaches a pre-determined value an alarm signal is operated and the slow turning rotary valve stops, indicating the location of the overheating.

Normal oil particles as spray are precipitated in the sampling tubes and drain back into the crankcase.

COOLING OF DIESEL ENGINES

The working temperatures of the gases inside the cylinders of diesel engines are very high, therefore the parts in close proximity to the combustion space must be cooled to prevent the metal from overheating. These parts are the cylinder head, cylinder and piston, and sometimes the exhaust valve. Fresh water is the most common cooling medium employed.

The cylinder is composed of a liner within a jacket, the liner is fitted at the top, with a water seal at the bottom and leaving an annular space between liner and jacket for the circulation of the water. The cylinder head is cast in box form with water passages running through it between the valve housings.

PISTON COOLING

The cooling medium for the pistons varies with different engines, it may be fresh water, distilled water, or lubricating oil. In some opposed piston engines lubricating oil is used for the bottom pistons and distilled water for the top, the lubricating oil being taken from the lubrication system. The cooling medium is led to and from the pistons by a system of telescopic piping, this consists of two pipes fixed to each piston, one for supply and one for return, these pipes oscillate in trombone fashion in larger diameter stationary pipes which are fitted with glands at entry to prevent loss. A sight glass is included in the return pipe so that the flow of cooling medium can be observed. A means of temperature measurement is incorporated in supply and return pipes. The temperature of the cooling medium is controlled by passing it through a cooler, usually circulated by sea water, before returning it to the reservoir. The quantity of cooling medium in the reservoir is noted regularly as a check against leakages in the system.

LUBRICATION

A forced lubrication system is always employed. The lubricating oil is fed under pressure by means of a pump or gravity tank to the main lubricating oil supply line, from a sump, passing through filters on the way. Pipes lead the oil from the main line to each main bearing and cam shaft bearing. A hole is drilled through the centre of the crank shaft, through the crank webs and crank pins which allows the oil to flow from the main bearings to the crank pins thence up through a hole in the connecting rod (or pipes strapped to the connecting rod) to the crosshead and guides.

The lubrication of the inside wall of the cylinder liner on which the piston rings rub, may be adequately effected by the oil mist thrown up from the cranks, or the oil may be pumped directly into the cylinder through two or more points, timed carefully to inject a few drops on the piston rings at the moment the piston is passing the oil holes.

The lubricating oil carries away a great deal of the heat generated by friction at the various bearings and therefore must be cooled by passing it through a cooler (which is circulated by sea water) at some stage of the circuit between leaving the sump and returning to the sump.

WARMING THROUGH

Some diesel engines are warmed through prior to starting. Although this may not be essential in some engines it is still good practice because, in addition to making starting easier, it minimises corrosion, wear, thermal shock and damage due to unequal expansion.

Warming through is done by heating the cooling systems. The cooling water storage tank is fitted with a heating element, usually steam, and the water is gradually heated to about 70°C while it is circulated through the engine (cylinder jackets, pistons, cylinder heads, etc.) by an independently driven pump.

In engines burning high viscosity fuel (heavy oil), the fuel valves and pipe lines are also warmed through before starting, the final warming-through temperature depending upon the oil viscosity required. The cooling circuit for the fuel valves, which is separate from the cylinder jacket system, may use water or fine mineral oil, and this is fitted with a heating element. Some oil pipe lines may be steam jacketed or have a small bore heating line wrapped spirally around them to heat the cold fuel oil standing in the pipes. A small auxiliary pump may also be used to circulate part of the fuel system with pre-heated oil. Failure to pre-heat the heavy oil and system could result in the engine failing to start.

STARTING

As previously explained, the ignition of the fuel in diesel engines is caused by the heat of compression of the air previously admitted into the cylinders, thus, for the engine to begin firing, air must first be drawn or pumped into the cylinder and this air must be compressed by the upward movement of the piston to obtain the high temperature necessary to burn the fuel when it is injected. Hence the engine must be driven for a few revolutions by some outside source before allowing the fuel into the cylinders. In heavy marine engines the practice is to drive the engine on compressed air which has previously been stored (at pressures ranging from 20 to 40 bar depending upon the type of engine) in starting-air reservoirs.

The compressed air is admitted to each cylinder through a starting-air valve when the piston has just passed its top centre and commencing what will be its power stroke and remaining open until the piston has travelled part of that stroke. The period of opening depends upon the number of cylinders and whether it is a four-stroke or two-stroke engine. When the starting-air valve closes on one cylinder, another starting-air valve has already opened on another cylinder whose piston has just commenced its downward stroke, when this valve closes another valve on another cylinder has already opened and thus, no matter in what position the engine stops there will always be at least one of the cylinders with its starting-air valve open to admit compressed air to start the engine. When the engine attains sufficient speed, the fuel pumps and valves are brought into operation and the starting-air valves put out of commission.

The reservoirs are pumped up by two-stage or three-stage compressors. In a two-stage compressor, atmospheric air is taken into the low pressure cylinder where it is compressed and discharged through an intercooler into the high pressure cylinder; it is now compressed to its final pressure and passed through an aftercooler into the air reservoirs. Sufficient air is stored to enable a minimum of about 14 starts of the engine without replenishing from the compressor.

SPEED CONTROL

The speed of the engine is controlled by varying the quantity of fuel injected into the cylinders, from zero quantity for stop to maximum quantity for full speed. This is commonly effected by varying the discharge period of the oil from the fuel pump.

REVERSING

There are two cams for each valve, one is set so that its peak will lift the valve rocking lever to open the valve and keep it open for the correct period of the cycle when the engine is running in one direction; the other cam alongside is set for running in the opposite direction. The reversing mechanism is therefore arranged to bring either ahead or astern cams into line with the valve rocking lever as required.

The action of one type of reversing gear is first to lift the rocking levers clear of the cams, slide the cam shaft along so that the opposite cam comes into line and then return the rocking lever to its working position.

SAFETY DEVICES AND ALARMS.

Some of these items are: –

Turning gear interlock to prevent starting of the engine with the turning gear in.

Governor to control the speed of the engine fitted with an overspeed trip.

Cylinder relief valves to prevent over-pressurising the individual cylinders.

Wrong way interlock and alarm, to prevent the engine going in the opposite direction to that demanded.

Alarms would be given for lubricating oil and cooling medium pressure and temperature.

STARTING AN AIR COMPRESSOR

Starting air is required for main and auxiliary diesel engines. To start an air compressor, open all air drains fitted to the air lines after the coolers. Ensure cooling water is on and open all valves on the suction and discharge sides of the compressor through to the air bottles. Start the prime mover that drives the air compressor, generally an electric motor with speed control. When the compressor is up to speed and air is blowing freely from the drains close the drains. At intervals these drains may have to be opened. When the air bottle is fully charged, open the drains, stop the compressor and close all valves.

STARTING A DIESEL GENERATOR.

This is similar to starting a main diesel engine. Check that cooling water to oil coolers is flowing. Check fuel oil and and lubricating oil supply. Turn the generator over with starting air until it reaches the desired speed then switch from air to fuel. The diesel will now run up to a governed speed without load.

At the electrical switchboard close the current circuit breaker for the generator and then gradually increase the electrical load until a balance is obtained between it and the generators already in operation.

HEAT BALANCE AND EFFICIENCY

The efficiency of different types of diesel engines varies but an average heat balance is as follows. Taking 100% as the heat given out by the fuel when it is burned inside the cylinders, about 40% of this is converted into work in the cylinders as indicated power (this percentage is the indicated thermal efficiency).

The remaining 60% of the heat in the fuel is divided into losses, about 28% of the heat is carried away by the cooling water, about 30% passes away in the exhaust gases, and the remainder, about 2% is lost by radiation.

Assuming the mechanical efficiency of the engine as 80%, the useful power at the crank shaft (brake power) represents 0.8 of 40% = 32% of the heat in the fuel, this percentage being the brake thermal efficiency; the other 8% represents friction losses.

Some of the heat in the exhaust gases may be recovered by passing them through an exhaust gas boiler (see Chapter 3), the steam generated being used for driving auxiliaries. More heat can be extracted out of the exhaust gases from a four-stroke engine than a two-stroke because of the higher temperature of the exhaust.

MEAN EFFECTIVE PRESSURE AND POWER

The indicator diagrams shown in Figs. 30 and 34 are examples of practical pV diagrams taken off engines by means of an engine indicator, the areas of these indicator diagrams represent the work done per cycle.

ENGINE INDICATOR. An engine indicator consists of a small bore cylinder containing a short stroke piston which is subjected to the same varying pressure that takes place inside the engine cylinder during one cycle of operations. This is done by connecting the indicator cylinder to the top of the engine cylinder, the gas pressure pushes the indicator piston up against the resistance of a spring, a choice of specially scaled springs of different stiffness being available to suit the operating pressures within the cylinder and a reasonable height of diagram.

A spindle connects the indicator piston to a system of small levers designed to produce a vertical straight-line motion at the pencil on the end of the pencil lever, parallel (but magnified about six times) to the motion of the indicator piston. The "pencil" is often a brass point, or stylus, this is brought to press lightly on specially prepared indicator paper which is wrapped around a cylindrical drum and clipped to it. The drum, which has a built-in recoil spring, is actuated in a semi-rotary manner by a cord wrapped around a groove in the bottom of it; the cord, passing over a guide pulley, is

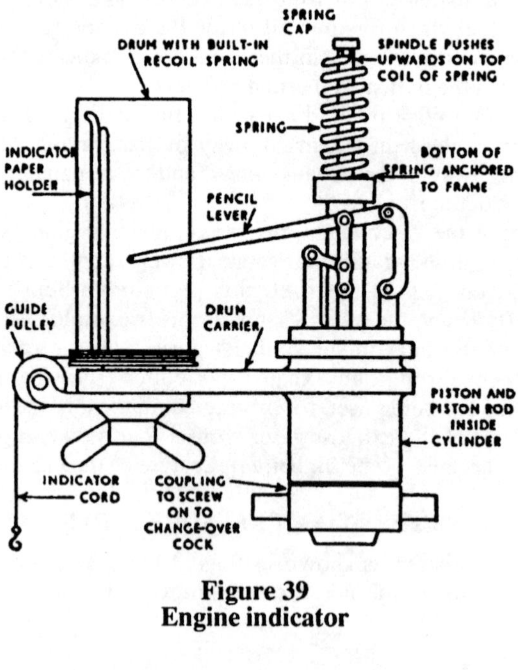

Figure 39
Engine indicator

attached by a hook at its lower end to a reduction lever system from the engine crosshead. Instead of the lever system from the crosshead, many engines are fitted with a special cam and tappet gear to reproduce the stroke of the engine piston to a small scale. The drum therefore turns part of a revolution when the engine piston moves down, and turns back again when the engine piston moves up, thus the pencil or stylus on the end of the indicator lever draws a diagram which is a record of the pressure in the engine cylinder during one complete cycle.

Fig. 39 shows a Maihak indicator which is suitable for taking indicator diagrams off internal combustion engines up to rotational speeds of about 300 rev/min. In this type, the pressure scale spring is anchored at its bottom end to the framework, and the top of the piston spindle bears upwards on the top coil of the spring, the upward motion of the indicator piston thus stretches the spring.

MEAN EFFECTIVE PRESSURE. Consider the two-stroke diesel engine indicator diagram shown in Fig. 40. The positive work done in one cycle of operations *by* the gas during the burning period and expansion of the gas is shown by the shaded area of Fig. 40b. The work done *on* the air during the compression period, representing negative work done by the engine is shown by the shaded area of Fig. 40c. Hence the net useful work done in one cycle is the difference between positive and negative work and represented by the actual diagram of Fig. 40a. Therefore, if the area of the indicator diagram is divided by its length, the average height is obtained which, to scale, is the average or mean pressure effectively pushing the piston forward and transmitting useful energy to the crank during one cycle. This, expressed in N/m² or a suitable multiple of the basic pressure unit, is termed the *indicated mean effective*

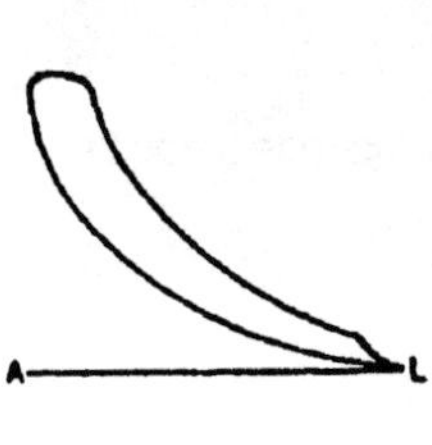

Figure 40a

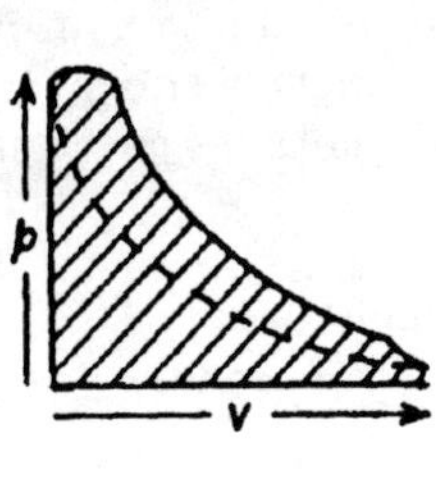

Figure 40b

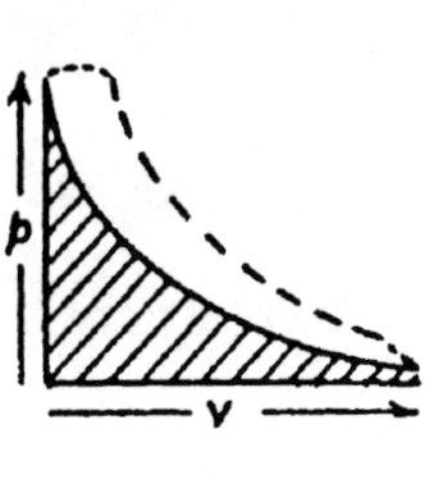

figure 40c

pressure.

The area of the diagram is usually measured by a planimeter. If the area is measured in mm^2 then dividing this by the length in mm gives the mean height in mm. The mean height in mm is now multiplied by the pressure scale of the indicator spring in N/m^2 per mm to obtain the indicated mean effective pressure in N/m^2. The usual convenient multiples of N/m^2 for such pressures are kN/m^2 and bars, and the spring may be graduated in either of these units.

If a planimeter is not to hand, the mean height of the indicator diagram may be obtained by the application of the mid-ordinate rule.

INDICATED POWER. Power is the rate of doing work, that is, the quantity of work done in a given time. The basic unit of power is the *watt* [W] which is equal to the rate of one joule of work being done every second. In symbols:

$$1 \text{ W} = 1 \text{ J/s} = 1 \text{ Nm/s}$$

The watt is a small unit and only suitable for expressing the power of small machines. For normal powers in marine engineering, mechanical, electrical or hydraulic, the kilowatt [kW] is usually a more convenient size, and heavy powers may be expressed in megawatts [MW]. The familiar unit of horse-power is now gradually becoming obsolete, powers of all engines, pumps, motors, compressors, etc., will now be measured in multiples of the watt. One horsepower is equal to 745.7 W or 0.7457 kW, hence, as an example, an engine previously rated as 14,000 hp is now expressed as 10,440 kW or 10.44 MW.

Let p_m = mean effective pressure [N/m^{2}1
A = area of piston [m^2]
L = length of stroke [m]
n = number of power strokes per second

then,

Average force [N] on piston
$= p_m \times A$ newtons

Work done [J] in one power stroke
$= p_m \times A \times L$ newton-metres = joules

Work per second [J/s = W]
$= p_m \times A \times L \times n$ watts of power

therefore,
Indicated power $= p_m ALn$

This is the power indicated in one cylinder. The total power of a multi-cylinder engine is that multiplied by the number of cylinders, if the mean effective pressure is the same for all cylinders.

Note that when the mean effective pressure is in N/m^2 the power obtained by the above expression is in watts. If the mean effective pressure in kN/m^2 is inserted, the result will be the power in kW, and this is usually more convenient.

The value of *n*, the number of power strokes per second, depends upon the working cycle of the engine (two-stroke or four-stroke) and its rotational speed.

In the four-stroke cycle, there is one power stroke in every four strokes, that is, one power stroke in every two revolutions, hence,

$$n = \text{rev/s} \div 2$$

In the two-stroke cycle, there is one power stroke in every two strokes, that is, one power stroke in every revolution, thus,

$$n = \text{rev/s}$$

Example. The area of an indicator diagram taken off one cylinder of a four-cylinder, four-stroke, internal combustion engine is $378mm^2$, the length is 70mm, and the indicator spring scale is 1mm = 1 bar. The diameter of the cylinders is 250mm, stroke 300mm, and rotational speed 5 rev/s. Calculate the indicated power of the engine assuming all cylinders develop equal power.

Mean height of diagram = area ÷ length
$= 378 \div 70 = 5.4$mm

Indicated p_m = mean height × spring scale
$= 5.4 \times 1 = 5.4$ bar

$5.4 \text{ bar} \times 10^2 = 540 \text{ kN/m}^2$
$n = \text{rev/s} \div 2$
$= 5 \div 2 = 2.5$

$$\begin{aligned}\text{Indicated power} &= p_m ALn\\ &= 540 \times 0.7854 \times 0.25^2 \times 0.3 \times 2.5\\ &= 19.87 \text{ kW per cylinder}\end{aligned}$$

Total power for four cylinders

$$= 4 \times 19.87 = 79.48 \text{ kW}$$ Ans.

BRAKE POWER AND MECHANICAL EFFICIENCY

Power is absorbed in overcoming frictional resistances at the various rubbing surfaces of the engine, such as at the piston rings, crosshead, crank and shaft bearings, therefore only part of the *indicated power* (ip) developed in the cylinders is transmitted as useful power at the engine shaft. The power absorbed in overcoming friction is termed the *friction power* (fp). The power available at the shaft is termed *shaft power* (sp) or, as this is measured by means of a brake it is also called *brake power* (bp).

$$\text{Brake power} = \text{indicated power} - \text{friction power}$$

The *mechanical efficiency* is the ratio of the brake power to the indicated power:

$$\text{Mechanical efficiency} = \frac{\text{brake power}}{\text{indicated power}}$$

Since the brake power is always less than the indicated power, the above expresses the mechanical efficiency as a fraction less than unity. It is common practice to state the efficiency as a percentage, by multiplying the fraction by 100.

Brake power is measured by applying a resisting torque as a brake on the shaft, the heat generated by the friction at the brake being transferred to and carried away by circulating water.

Let F = resisting force of brake, in newtons, applied at a radius of R metres when the rotational speed is in revolutions per second, then:

$$\begin{aligned}\text{Work absorbed per revolution [Nm = J]}&\\ &= \text{force [N] x circumference [m]}\\ &= F \times 2\pi R\end{aligned}$$

$$\begin{aligned}\text{Work absorbed per second [J/s]} &= \text{power absorbed [W]}\\ &= F \times 2\pi R \times \text{rev/s}\\ F \times R &= \text{torque in Nm} = T,\end{aligned}$$

$\therefore$ brake power $= T \times 2\pi \times$ rev/s

$2\pi \times$ rev/s $=$ angular velocity in radians/second

$= \omega$

$\therefore$ brake power $= T\omega$

Propeller Power. The power given out by the propeller will be less than the shaft power due to slip and turbulence.

PETROL ENGINES

Engines which run with petrol (or paraffin) as the fuel are sometimes termed "light oil" engines. The main difference between the average petrol engine and the diesel engine is that the petrol engine takes in a charge of air and petrol vapour, this explosive mixture is compressed and ignited by an electric spark; whereas in the diesel engine the cylinder is charged with air only so that only pure air is compressed and the fuel is injected at the moment ignition and burning of the fuel is required, ignition being caused solely by the heat of the compressed air.

When the air is compressed in a diesel engine there is no possibility of firing before the fuel is injected. In a petrol engine, when an explosive mixture of petrol and air is compressed, there is a danger of the mixture firing spontaneously due to the heat of compression alone and before the electric spark occurs, therefore the ratio of compression must be limited to prevent this. The ratio of compression in diesel engines can be high, such as 12 to 1 and upwards, whereas the ratio of compression in spark-ignition petrol engines is lower, usually less than 10 to 1.

Petrol engines usually work on the "constant volume" cycle, that is, when combustion takes place there is theoretically no change in volume but a considerable increase in pressure.

CHAPTER SIX

PUMPS AND AUXILIARIES

As an example of the number, names and connections of pumps installed in a ship's engine-room, the table on the following page is given as being typical for a steamship. A suction valve connection to a given service is denoted by the letter S and a discharge valve connection is denoted by the letter D.

Special service pumps such as the oil fuel pressure pumps and transfer pumps in oil fired boiler installations, lubricating oil pumps for forced lubrication of engines, brine and circulating pumps for refrigerators, hydraulic pumps for hydraulically operated water-tight doors, cargo pumps and so on, have been omitted to avoid a lengthy and complicated list.

TYPES OF PUMPS

The pumps employed on board ship can be divided into two main categories (i) displacement pumps, (ii) centrifugal pumps, and in each category there are several types.

Displacement pumps are those where the volume of the pump chamber is alternately increased to draw the liquid in from the suction pipe and then decreased to force the liquid out into the delivery pipe. This may be done by either a reciprocating motion of a piston, ram or bucket within a cylinder, or by a rotary motion of specially designed vanes, gears or rotors.

Centrifugal pumps are those wherein an impeller rotating at high speed throws the liquid by centrifugal force from the centre (to which the suction is connected) radially outwards to the periphery of the impeller where the liquid is discharged through the delivery outlet.

In general, displacement pumps of the reciprocating type are suitable for delivering small quantities at high pressures, rotary

PUMP CONNECTIONS

	BOILER FEED PUMPS	GENERAL SERVICE PUMP	CONDENSER CIRCULATING PUMP	BALLAST PUMP	BILGE PUMP	CONDENSER EXTRACTION PUMP	FRESH WATER PUMP	SANITARY PUMP	EMERGENCY FIRE PUMP
SEA		S	S	S	S			S	S
HOTWELL OR FEED TANK	S	S				D			
BALLAST TANKS		SD		SD					
FRESH WATER D.B. TANKS		S		S			S		
ENGINE ROOM BILGE		S	S	S	S				
MAIN BILGE LINE		S		S	S				
CONDENSER CONDENSATE						S			
BOILERS	D	D							
OVERBOARD		D		D	D				
CONDENSER WATER-BOX			D	D					
WASH-DECK AND FIRE-SERVICE		D		D	D			D	D
SANITARY TANKS		D		D				D	
FRESH WATER HEAD TANKS		D		D			D		
FORE AND AFT PEAKS		D		SD					

displacement pumps are used for moderate quantities at moderate pressures, and centrifugal pumps are more suitable for large quantities at low pressures. Centrifugal pumps however, can be designed with a number of impellers in series to attain a high final delivery pressure and, because of the simplicity in direct drive from an electric motor, centrifugal pumps are now gradually superseding reciprocating pumps for most duties.

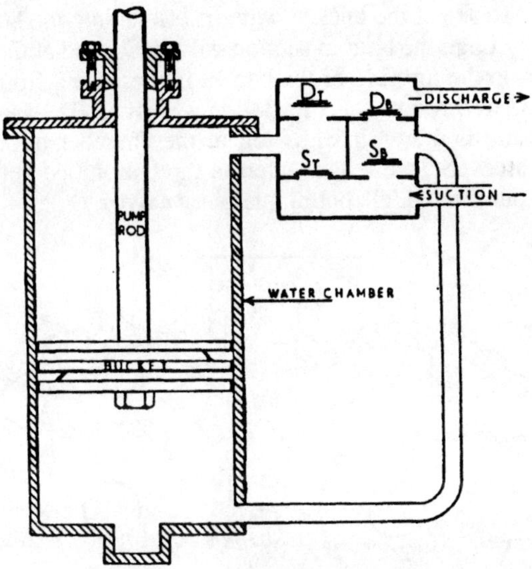

Figure 41
Double-acting piston pump

DOUBLE ACTING PISTON PUMP. This is a common example of a reciprocating displacement pump. The reciprocating motion may be obtained through a connecting rod and crank mechanism from an electric motor drive or directly from a steam piston in its own cylinder. In the latter case there is a specially designed slide and/or piston valve actuated by valve gear, which distributes steam alternately to the top and bottom side of the steam piston to give it its reciprocating motion. The steam piston is connected directly to the water piston by piston rod, crosshead and pump rod. To distinguish between the steam and the water cylinders of the pump it is usual to refer to the water cylinder as the pump "chamber", and the water piston as the "bucket". It is a double-acting pump which means that water is discharged from both the top and bottom sides of the bucket, therefore water is discharged on every stroke. Each end of the water chamber has its own set of suction and delivery valves. The two sets of suction and delivery valves are housed side by side in the one casting, one set being connected by a port leading into the top of the chamber and the other set by a port leading to the chamber bottom.

Fig. 41 shows the pump chamber and valves diagrammatically. On the up-stroke of the bucket, water is drawn into the bottom of the chamber through the bottom suction valves, S_B, and at the same time the water on the top side of the bucket is forced out from the top of the chamber through the top delivery valves, D_T. On the down stroke, water is drawn into the top of the chamber through the top suction valves, S_T, while the water on the bottom side of the bucket is forced out through the bottom delivery valves, D_B.

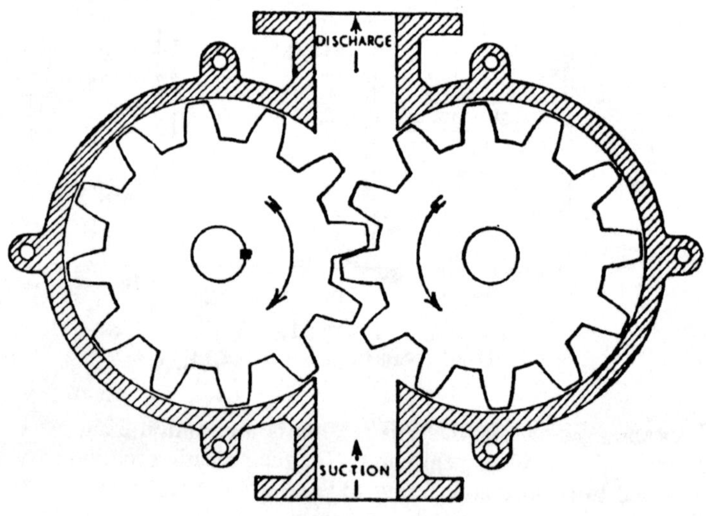

Figure 42
Gear wheel pump

GEAR WHEEL PUMP. Fig. 42 illustrates a gear wheel pump which is an example of a rotary displacement pump. These are used extensively for pumping lubricating oil and fuel oil. Note the direction of rotation of the gear wheels in relation to the flow of oil through the suction and delivery branches. As each tooth in one wheel leaves its corresponding space in the other wheel, a partial vacuum is created in that space into which the oil flows, the oil is then carried around in the space almost through 360 degrees when a tooth entering that space forces the oil out into the discharge line.

In gear wheel pumps which are driven off the main engine shaft, the oil is required to flow always in the same direction whether the engine is running ahead or astern. This is achieved by adding a pair of non-return suction and delivery valves at each side with communicating ports to the suction and delivery branches.

CENTRIFUGAL PUMP. This is a rotary pump which works on the principle of centrifugal force, that is, that outward radial force set up by a mass rotated in a circular path due to its natural tendency to fly off at a tangent to the circular path and travel in a straight line.

The pump consists of a rotating impeller within a stationary casing. The impeller is like a hollow disc wheel with internal curved vanes, mounted on a shaft which is driven by an electric motor, steam engine or turbine, or other prime mover. Openings in the sides of the impeller near the shaft communicate with the suction branch, water (or oil, etc.) enters the rotating impeller through these ports, and due to the circular motion given to the water it is thrown by centrifugal force to the open periphery of the impeller which it leaves tangentially and enters the space between the outer circumference of the impeller and the casing and directed to the outlet branch. See Fig. 43.

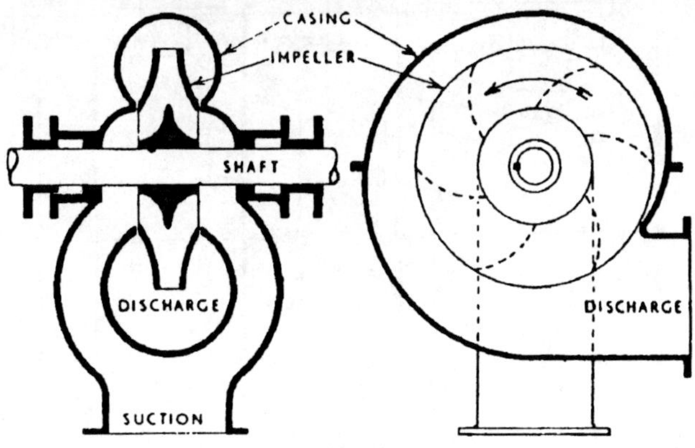

Figure 43
Centrifugal pump

Unlike the reciprocating displacement pump, the centrifugal pump does not have a positive suction action and must be primed by flooding before it will draw water from a lower level. It is therefore employed mainly where the suction is submerged or the lift is very small. Centrifugal pumps will only pump in the one designed direction of rotation.

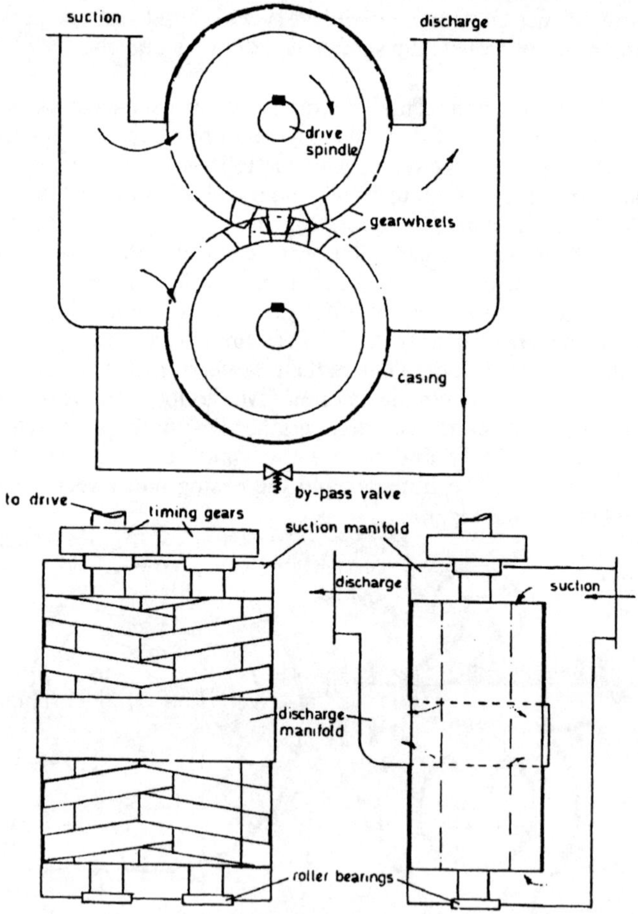

Figure 44
Screw displacement pump

The drive for these pumps is most often directly from an electric motor but can be from an auxiliary turbine. In the latter case the prefix *turbo* is adopted, for example, "turbo feed pump."

SCREW DISPLACEMENT PUMP. This type of pump is capable of handling intermittent fluid supply and hence is well suited to tank draining and to lubricating oil systems for diesel engines when the vessel is rolling.

Its operating principle is that of the ancient Archemedian Screw used for irrigation purposes. The fluid could be likened to a nut on a rotating screw thread. The nut is prevented from rotating and must therefore move linearly along the thread.

Fig. 44 shows diagrammatically a screw displacement pump. Fluid enters the outer suction manifold and passes through the meshing screws, which are gear driven from a motor, to the central discharge manifold.

VARIABLE DELIVERY PUMP. This type of displacement pump is used in hydraulic systems, *e.g*. steering gears, stabilisers etc.

Rotating pistons and cylinders are driven at constant speed by an electronic motor. One type, the Hele-Shaw, has the pistons and cylinders in the same plane, but their axes of rotation can be adjusted and made electronic so that the pump can deliver fluid in variable and can alternate between suction and discharge depending upon the eccentricity of the axes. *i.e*. a suction becomes a discharge port and vice-versa.

BILGE PUMPING ARRANGEMENT

Fig. 45 is a line diagram of a typical bilge suction arrangement where the bilge pump, ballast pump or general service pump can be used for pumping out any bilge.

The distribution valve chests are situated in the engine room to enable any bilge to be pumped out by the watch-keeping engineer. All bilge suction valves are of the screw-down non-return type to prevent water flowing back and flooding the bilges, the operating hand wheels are labelled by engraved brass plates as to which bilge the pipe runs into. A mud-box is fitted on the bilge suction of each pump and the open end of every bilge pipe in the bilges is enclosed in a strainer box.

In addition to this system there is the engine-room bilge injection connected to the condenser circulating pump as shown in Fig. 52,

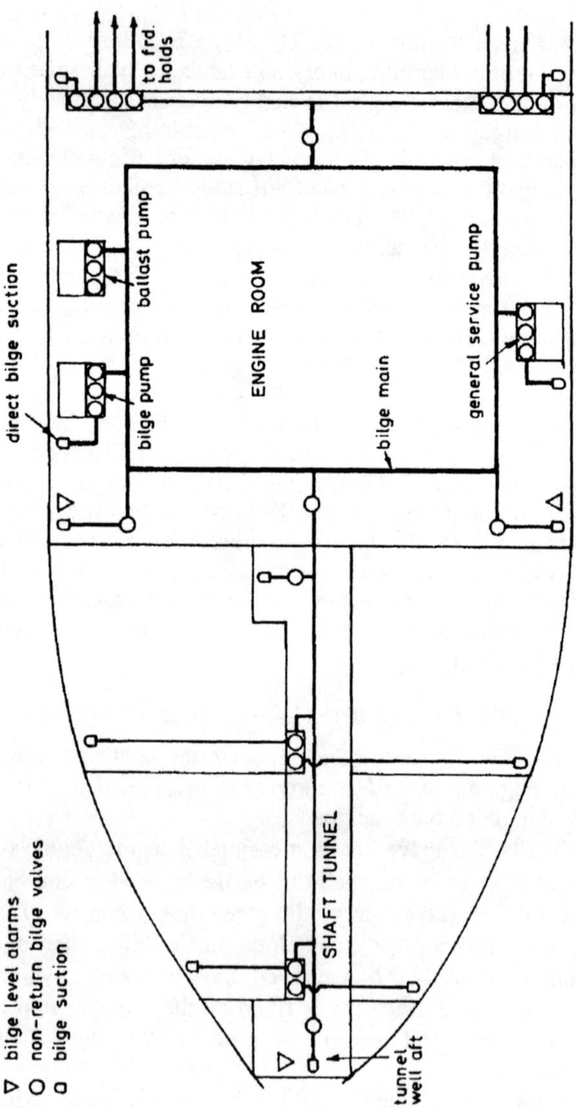

Figure 45
Bilge pumping arrangement

and the emergency bilge pump as described below.

EMERGENCY BILGE PUMP. This, as its name implies, is used in an emergency for pumping water out of the ship when a compartment is flooding due most likely to hull damage.

It is a self-contained unit consisting of a centrifugal pump to deal with the water, reciprocating or rotary air pumps to rid the water suction of air to allow priming of the centrifugal pump, and an electric motor to drive the pumps. The drive shaft is vertical and the electric motor is above the pumps, the motor being enclosed in an air bell to protect it from being flooded when the compartment is full of water, thus the system continues to work when the unit is completely submerged.

The electric supply is taken from the ship's emergency electrical circuit and the unit can be operated by remote control.

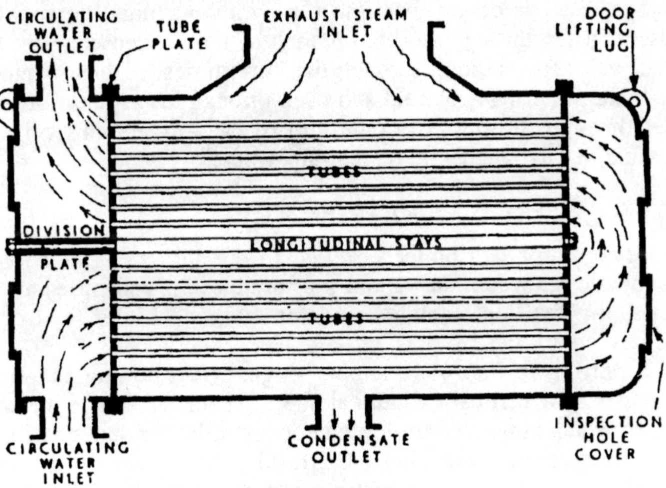

Figure 46
Surface condenser

AUXILIARY CONDENSER

All ships' condensers are of the surface type and vary in shape and design. A surface condenser commonly used in conjunction with steam auxiliaries is a two-pass type consisting of two nests of sea water cooled tubes as shown in Fig. 46. It consists of a shell constructed of mild steel plates welded together, either cylindrical or pear-shaped in section, with a flange around the periphery of each end to which is bolted a brass tube plate. The tubes are about 19mm diameter and 1.5mm thick, and may be made of brass or a copper-nickel alloy, they are roller expanded into the tube plate at the inlet end and fitted with a simple leak proof gland, which allows for expansion and contraction, at the outlet end. A cast iron or coated prefabricated steel water chamber is bolted to the tube plate at each end. One of these water chambers is formed simply by the convex shape of the condenser door, the other consists of a box with a sea water inlet branch at the bottom, a horizontal division plate across the centre, a sea water outlet branch at the top, and a separate door bolted on the outside. A large opening in the top of the condenser shell forms the inlet branch of the exhaust steam and a small opening in the bottom forms the branch through which the condensate and air is drawn.

Sea water is drawn from the sea by a centrifugal pump which discharges it through the inlet branch of the condenser water box. The water flows along through the bottom nest of tubes, into the opposite water chamber, up and back through the top nest of tubes, out through the outlet branch and overboard via an overboard discharge valve higher up on the ship's side.

EVAPORATORS

An evaporator is a boiler installed in the ship's engine room to evaporate sea water, the vapour thus produced is condensed and the resultant distilled water is used for make-up boiler feed or for domestic purposes.

Referring to Fig. 47, a simple single-effect boiling evaporator consists of a vertical cylindrical shell, the lower half is the water space which houses steam-heated copper coils, the upper half is the vapour space and contains a baffle plate to throw off any water particles which might rise with the vapour. One end of each copper coil is connected to a steam header which is supplied with steam either from the auxiliary steam line or the exhaust range, the other end of each coil is connected to an exhaust header with a drain at the bottom. The evaporator illustrated shows an arrangement whereby all the exhaust steam and condensed water from the upper coils finally pass through the bottom coil to ensure that all the heating steam has been condensed and thereby given up all its latent heat before leaving by the drain as water, this is a feature of Messrs. Weir's evaporator. A large hinged door is provided in way of the coils and also a hand-hole near the bottom of the shell.

Sea water is pumped into the evaporator, and the hot coils cause the sea water to boil, the vapour given off passes through the vapour valve at the top and is led to either the condenser where it is condensed to make extra feed water for the boilers, or to a distiller where it is condensed as a make-up for the domestic tanks. In some installations the vapour may be passed to the heating side of a feed heater where its latent heat as well as some of its sensible heat can be given up to the feed water.

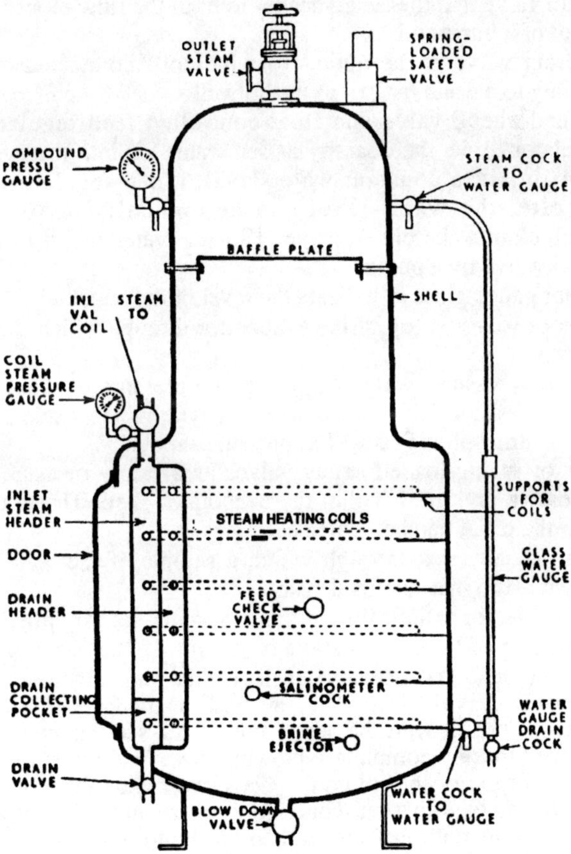

Figure 47

As the sea water boils and passes away as vapour, the salt and other solids are left behind which causes an increase in the density of the water in the evaporator, therefore to maintain a pre-determined steady low density the evaporator is run with a constant blow-down through a brine ejector. Some of the solid matter deposits as scale on the heating coils, this is cleaned off at regular intervals.

The fittings on this evaporator are:

A steam valve to the coils, a steam pressure gauge to indicate the pressure of the heating steam and a nozzle plate between the steam valve and the steam header to limit the flow of steam in the event of a burst coil.

A drain valve at the outlet from the coils connected to a pipe leading to a drain tank or to the hotwell.

A feed check valve and float-controlled feed regulator. The regulator limits the quantity of sea water fed into the evaporator to maintain a constant water-level, if the regulator fails to operate, the water-level can be manually controlled by manipulating the check valve. The sea water is fed to the feed check valve by a pump.

Water gauge glass to indicate the level of the water.

Vapour valve on top of the evaporator through which the vapour passes.

Vapour pressure gauge to indicate the pressure of the vapour in the evaporator. This is a compound type which reads pressures below atmospheric as well as pressures above.

Pair of spring-loaded safety valves as a safety measure against excessive pressures within the evaporator shell. These are fitted with the usual hand operated easing gear.

Salinometer cock through which a sample of the water can be drawn off to measure its density.

Brine ejector which incorporates a water jet (supplied from a pump) to induce a constant blow-out of the water from the evaporator so that a steady density can be maintained.

Blow-down valve fitted to the bottom of the evaporator and communicating with the sea. Some of the scale may be cracked off the tubes by completely blowing down and allowing the cold sea water to rush back into the evaporator while the tubes are still hot, the sudden thermal contraction of the tubes causes some of the scale to flake off and fall to the bottom of the evaporator. This is usually done at regular intervals.

FLASH EVAPORATOR. This is another type of evaporator in

common use. The saturation temperature of water or steam is the temperature at which water evaporates or steam condenses and this depends upon the pressure, the higher the pressure the higher the saturation temperature, the lower the pressure the lower the saturation temperature. Hence, if water under pressure is at its saturation temperature and its pressure is suddenly reduced, the temperature of the water must fall and in doing so releases some of its heat energy. The heat energy released is absorbed by a portion of the water which causes it to evaporate.

The process of quickly reducing the pressure of hot water to cause some of it to flash into steam vapour is termed *flash evaporation* and is the principle upon which this type of evaporator works.

The flash evaporator shown in Fig. 48 consists of a shell and base made of mild steel, internally coated with a bonded synthetic rubber to protect against corrosion, demister screen of knitted monel metal wire which scrubs the vapour of sea water droplets, a vapour condenser made up of aluminium brass tubes expanded into a naval brass tube plate, a heater which is supplied with steam or diesel engine cooling water, and three pumps, brine, air and distillate.

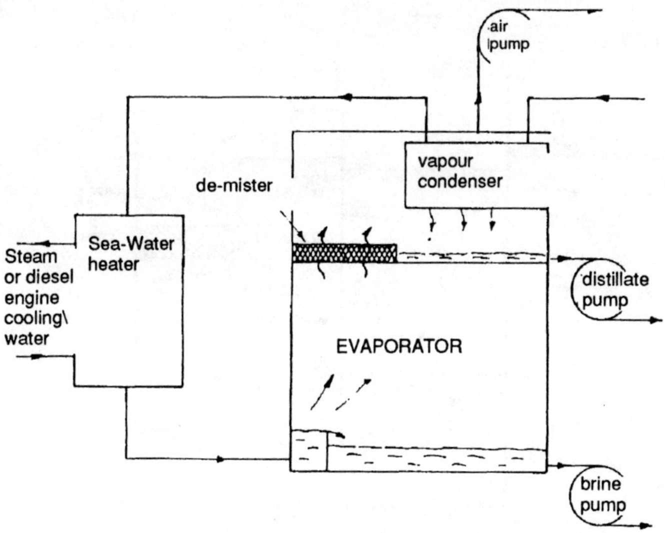

Figure 48
Single stage flash evaporator

To operate the evaporator, the air pump distillate and brine pumps are started. The heating steam is then turned on together with the sea water from the ship's mains (or alternatively from a separate pump used only for the evaporator). When the pre-heated slightly-pressurised sea water enters the evaporator shell some of the water flashes off into steam since the pressure inside the shell is very low. The vapour is condensed, collected and pumped out by the distillate pump. Any unevaporated sea water is pumped out by the brine pump.

Flash evaporation systems often comprise two or more stages. In a simple two-stage flash evaporating plant, pre-heated sea water is pumped under pressure into the first of two compartments which is maintained at a lower pressure, on being released some of the water flashes into steam vapour. The remainder of the sea water is now passed into the second compartment which is maintained at a further lower pressure and again, on pressure release, more water flashes off into steam.

Cooling water, circulating through diesel engines, absorbs heat which can be used in a fresh water generator. This provides an economy, as the heat in the cooling water would normally be exhausted to the sea, and is current common practice.

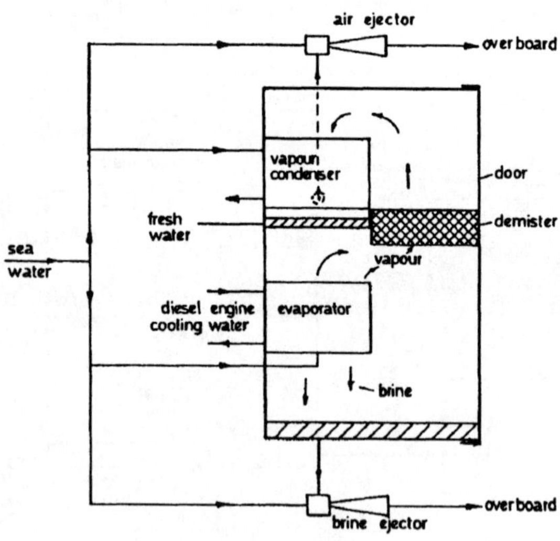

Figure 49
Fresh water generator

Fig. 49 illustrates diagrammatically a fresh water generator. The shell is fabricated steel protected internally from the corrosive effect of sea water by a bonded, vulcanised rubber coating. Evaporator and vapour condenser heat exchangers are either plate or tube type made of aluminium-brass, or possibly titanium in the case of the evaporator. The evaporator section operates in a similar way to that already described for Fig. 47 so it could be described as a submerged element vacuum evaporator. A sea water supplied air ejector removes non-condensible gases, which are liberated from the evaporated sea water, and produces in combination with the vapour condenser a low pressure inside the shell. Some of the sea water fed into the evaporator boils off at about 50°C because of the low pressure, the remainder is discharged as brine. Condensed vapour is removed by the fresh water pump and delivered to storage tanks.

WARNING: Due to the low temperature of operation in low pressure fresh water generators the made fresh water may be harmful to drink. Sterilisation temperature is about 80°C, however sodium hypochlorite (one part per million) may be used to chlorinate and hence sterilise the water. Other methods of sterilisation, electrolytic or use of ultra-violet light may be used but if the vessel is within certain coastal limits the sea water may contain harmful effluents from industrial zones and none of the sterilisation methods will give protection.

REVERSE OSMOSIS PLANT. Osmosis is a natural phenomena where solutions of different concentrations separated by a semi-permeable membrane will flow from the solution of low concentration to the higher. This process can be reversed by the application of high pressure.

Pre-treatment of the sea water feed is essential to minimise possible damage to the membrane. The pre-treatment consists of a duplicated filter unit. Sterilisation by chlorine, followed by a dechlorinator, or use of U.V. light. Then continuous chemical dosing to deal with organic materials,gases etc.

The pre-treated feed is pressurised by a pump up to as high as 1000 bar for delivery to the membranes. After the membranes the water is relatively salt free, but if the water is needed for high pressure boiler use it will require further de-ionisation treatment. This type of plant has been produced for ship-board use as some modern propulsion machinery has limited amounts of waste heat that can be used to produce fresh water by vapourisation. The plant has the advantage that it can be used when the machinery is shut down.

FEED HEATERS

A feed water heater is included in the feed system to heat the feed water to a temperature of 105°C and upwards. Feed heating increases the efficiency of the plant.

There are two main types of feed water heaters, (i) the *surface feed heater* which uses steam as the heating medium, and (ii) the *economiser* which uses the boiler flue gases as the heating medium. The surface feed heater is so called because it was formerly necessary to distinguish this type from the *contact feed heater* which also used steam as the heating medium but is not now fitted in modern installations.

THE SURFACE FEED HEATER is installed between the discharge side of the feed pump and the boilers. It consists of a cylindrical shell containing nests of tubes, the feed water flows into the inlet branch of the water header, down through one nest of tubes, up through another nest, and so on until it is discharged through the outlet branch of the heater. Steam enters the top, passes around the outsides of the tubes and out through the drain at the bottom, it passes out as water having given up its latent heat to the feed water through the walls of the tubes. The heating steam may be taken from the auxiliary exhaust line, or live steam may be bled off some part of the steam range. Two heaters may be installed in which case the first heater through which the feed water passes is heated by exhaust steam, and the second heater heated by live steam.

AN ECONOMISER is an extra feed water heater wherein the water, after leaving the normal feed heaters, is further heated by the combustion gases in the boiler uptake, as shown in Fig. 14, Chapter 3. The feed water enters at the top of the economiser, passes through the tubes and headers, and leaves at the bottom (which is the hottest region of the gases) to pass directly into the steam drum.

STEAM TRAP

On steam heating systems such as oil fuel heaters and tank steam heating coils, it is necessary that only the condensate passes on to the drain or feed tank, this ensures that the heating steam has given up all of its latent heat energy to the oil in the heaters. A steam trap is the usual device fitted to fulfil this operation.

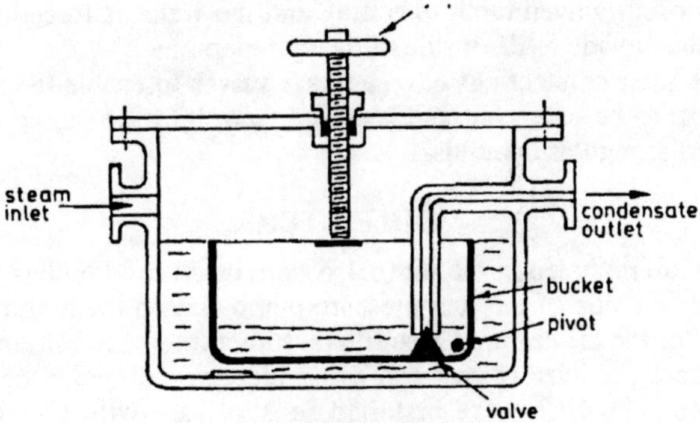

Figure 50
Steam trap

Fig. 50 shows the basic parts of a steam trap of the bucket type. The return line from the heaters carries a mixture of water and steam and this is connected to the steam inlet of the trap. As the water collects in the bucket, the buoyancy of the bucket is destroyed, it tips about the pivot and opens the valve. This allows the water condensate to be driven out under the steam pressure, the buoyancy of the bucket is then gradually restored and the valve closes to prevent the escape of steam.

To purge the system of air the by-pass is operated with the heating steam turned on, when all air has been expelled and steam issues from the condensate outlet, the by-pass is released and the trap left to operate automatically.

FEED WATER FILTERS

The function of a feed water filter is to extract oil and other impurities from the water before it is fed into the boiler. In steam reciprocating pumps the piston and valve rods are swabbed with oil, some of this is carried into the cylinders, finds its way into the condenser with the exhaust steam, and mixes in the condensate. Any small quantities of oil which might be carried into the boiler could become a great danger, it is a poor conductor of heat – worse than scale, and a patch of oil on a heating surface can cause over-heating of that part with consequent loss of strength and collapse. Oil must therefore be kept out of the boiler and the filter is in the feed line to perform that duty.

The filters are usually fitted in the discharge line of the feed pumps. A common type consists of perforated cylinders wrapped with closely woven cloth towelling and the water is forced through from the outside to the inside of the cylinders.

The filter connections have by-pass valves to enable the filtering medium to be taken out and renewed, and the filter boiled out and cleaned at regular intervals.

OIL FILTERS

Filters are included in the oil fuel system of oil fired boilers both on the suction side of the fuel pressure pump and on the high pressure side after the oil heater. These filters, sometimes called strainers, are to extract particles of grit and dirt which would choke the burner nozzles. The filters are installed in duplicate with change over valves so that one can be cleaned while the other is in use, pressure gauges are connected to the inlet and outlet sides of the filters so that any difference in pressure can be seen which would indicate a dirty filter. The filter housing contains a perforated steel cylinder covered by steel gauze through which the oil is forced, the mesh of the gauze on the hot filters being finer than that on the cold filters.

Sometimes Auto-Klean filters are fitted which have the advantage of being capable of being cleaned while in use. This is done by turning a handle on the top which actuates cleaning blades and dislodges dirt to fall to the bottom of the chamber, the chamber is cleaned out at any convenient time.

Similar filters are included in diesel fuel and lubricating lines and in turbine lubricating oil systems.

CENTRIFUGAL OIL SEPARATOR

When lubricating oil becomes contaminated with water, the water can be separated from the oil speedily and efficiently by a centrifugal oil separator. One example of contamination is in the lubricating oil system of a turbine installation. Steam from the glands condenses and finds its way through the bearings into the oil, and salt water leakage could occur in faulty oil coolers. Water, more especially sea water, has a bad effect on the lubricating quality of the oil and there is also the tendency to rusting of the steel parts.

The centrifugal oil separator, as its name suggests, works on the principle of centrifugal force. The mixture of oil and water is fed into the top of the separator through a strainer, flows down into a bowl rotating about its vertical axis at a very high speed (gear driven by an electric motor) and the water, being heavier than oil is thrown outwards at the bottom, up through an annular space into a stationary trap from which it flows through a spout into a sludge tank or the bilges. The purified oil passes upwards through another annulus and trap and is led back into the sump. In some separating systems, the oil and water mixture is heated before being admitted to the separator.

EMERGENCY GENERATOR

This is a most important auxiliary unit. It is for use under conditions of emergency when the main generators are out of action, to supply electrical energy for the essential emergency services. These include navigation lights, emergency lighting, communications, emergency bilge and fire pumps, and operation of water-tight doors.

The prime mover driving the generator is a completely independent self-contained compression-ignition internal-combustion engine with its own separate source of fuel supply, the fuel being of good quality light diesel oil which makes easy starting from cold.

Prior to starting, the cooling and lubricating systems are checked. The method of starting depends to a large extent on the size of the engine. For large units the engine is started by compressed air in a similar manner to the main engines, the air being stored in separate starting-air bottles. Usually, the turning of one handwheel opens the starting-air to the engine which runs the engine quickly up to working speed, turning the same handwheel further cuts the starting-air off and the fuel system on. Medium and small sized engines may be started by hand-cranking or using a hydraulic starter, starting cartridges may be fitted into holders in the combustion spaces to provide hot spots and hence assist the initial firing. Many small sized engines are started by electric motor in a similar manner to the starting mechanism of a motor car engine except that the battery supply is more often 24 volts, the batteries are kept fully charged by a trickle-charger.

Situated in the same compartment as the emergency generator is the switchboard which is connected to the emergency services mentioned above, and to the main switchboard in the engine room. When emergency power is required to be transferred from the main generating system to the emergency generating system, the voltage is adjusted to the desired value, emergency circuit breaker closed and the breaker from the main system opened. In the event of main electrical power failure in modern systems, the emergency generator is designed to start automatically and give emergency power without human intervention.

All emergency generators must be tested frequently and regularly by starting, running and switching-over the emergency circuits from main supply to emergency supply. The whole system must be maintained in perfect running order and always in complete starting readiness, for example, fuel supply tank(s) full, air at the correct pressure in the starting-air bottles, or fully charged batteries.

THE WINDLASS

The duty of the windlass is to lift the anchors and assist in warping the ship and therefore its size and power depend upon the masses of the anchors and cable, and full load hauling, which is governed by the size of the ship. It may be powered by a steam engine or electric motors.

The basic design is that of a double-purchase lifting machine consisting of a primary shaft, intermediate shaft, and two main half-shafts, with corresponding pinions and gear wheels as shown diagrammatically in Fig. 51. In the steam reciprocating engine driven type, the primary shaft of the windlass is the engine shaft of a two-cylinder steam engine; in the electrically driven windlass, the primary shaft is driven by worm and wormwheel through a worm shaft, from the electric motors.

The primary shaft carries a pinion which meshes with a gear wheel on the intermediate shaft, and two pinions on the intermediate shaft mesh with two main gear wheels, one on each main half shaft. Each main half shaft carries a cable-lifter which has snugs around its circumference of the size and pitch to suit the links of the cable.

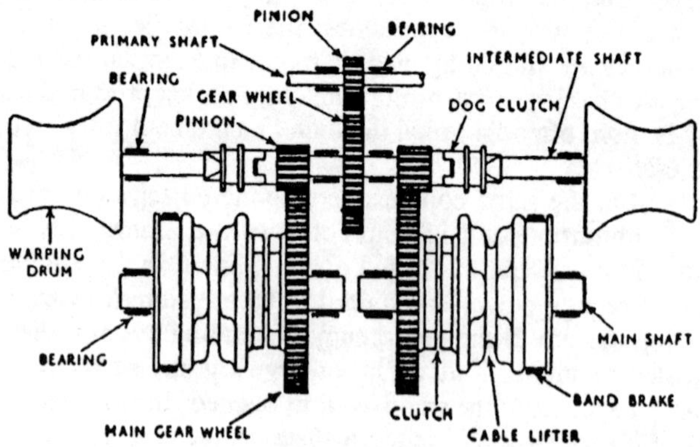

Figure 51
Windlass

The cable-lifters are not fixed on the shafts but are mounted freely to allow them to rotate independent of the shafts. A screw-operated steel band brake is fitted around a brake drum on the outer edge of the rim of the cable-lifter for controlling the speed of the cable when "paying out" and for locking it stationary when required. The power for hoisting is transmitted through a clutch formed by jaws on the side of the main gear wheel which fit a corresponding set of jaws on the side of the cable-lifter. The main gear wheel may be a sliding fit and keyed to its half shaft to allow it to be moved laterally into and out of gear, alternatively the gear wheel may be fixed on the shaft and the cable-lifter moved laterally to engage gear. A screwed control rod attached to a cod-piece riding in a groove in the boss of either the main gear wheel or the cable-lifter operates the clutch. Thus the two cable-lifters are entirely independent, the anchors may be lifted both at once, or separately, or one may be lifted while the other is being "let go."

Each end of the intermediate shaft is extended through a dog clutch to carry a warping drum. In the event of power failure, the windlass can be operated by hand gear consisting of a lever and pawl to act as a ratchet on the teeth of the intermediate gear wheel.

SHIPSIDE FITTINGS

Valves and cocks are fitted to openings in the hull of the ship to admit or discharge sea water and can be divided into two categories, sea water inlet and overboard discharge.

Sea inlets are situated well below the light load water-line so that they are completely submerged for all conditions of loading with allowance made for the rolling of the ship. In general, valves are fitted for medium and large quantities of water, cocks may be fitted for small quantities. The valves are most often of the globe type with the water admitted to the underside of the valve, they are of the screw-lift pattern with external screw, that is, the screw of the spindle operates through a bridge fixed on pillars fitted outside the valve cover, the glands can then be repacked without waiting for dry-docking and the valve can still be opened in the event of a broken valve spindle. Some medium sized valves may be of the gate type. The outer flange of the valves have a spigot which fits neatly into the opening in the ship's side. Where large and heavy valves are fitted, the shell is compensated for loss of strength caused by the large opening, by fitting a pad around the hole. The flanges are secured to the ship's side by brass bolts and nuts, the bolts being screwed through the ship's plate. The valves may be made of cast iron, cast steel, or gun-metal, with gun-metal fittings.

The openings in the ship's side are covered by grids to prevent marine bodies and other foreign matter entering and choking the inlet system. The slots in the grid are usually 12mm or less in width, the total area of opening is not less than 25% greater than the area of the valve and may be anything up to twice the area to allow for partial blockage. The grids are usually fitted with their slots in a horizontal direction, this gives some chance of clearing itself of foreign matter by the water stream of the ship's motion.

When the valves or grids are made of gun-metal, zinc rings or slabs are fitted around the openings to minimise corrosion of the ship's steel hull due to galvanic action.

At one time a hole was cut in the ship's side as near as practicable to every pump requiring a sea suction valve but it is more common in modern ships to fit a watertight steel box to take a number of, or all, the sea suction valves, thereby reducing the number of openings in the hull. The open side of this sea-box is shaped to fit the curvature of the lower part of the side of the hull. A grid is fitted over the opening of the box and the valves are mounted on its flat top and inner side.

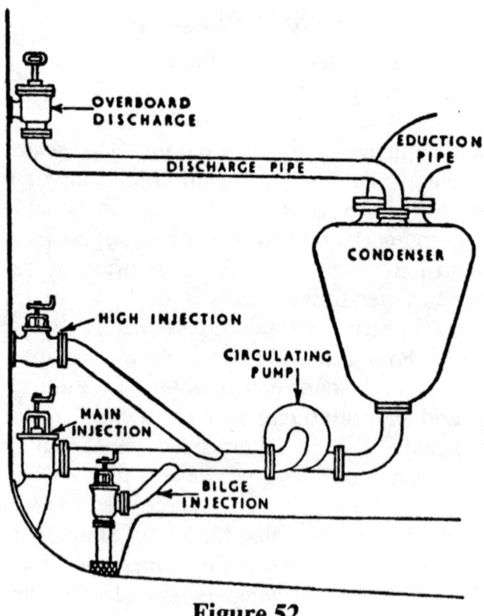

Figure 52
Condenser connections

In steam ships which include shallow water in their trade routes, separate high and low injections are fitted with a common sea water suction rail to the condenser so that the high injection can be used when the low position would be liable to pick up sand or mud (see Fig. 52).

On every dry-docking the opportunity is taken to fully inspect all valves, sea-boxes and other openings below the water-line. Valves are thoroughly overhauled, ground-in, gland packing and joints renewed, grids cleaned, sea-boxes scraped on the inside and coated with anti-corrosive paint, and new zinc anodes fitted where necessary.

Overboard discharge valves are mostly above the full-load water line. Non-return valves are fitted on small outlets but large valves, such as the steam condenser overboard discharge, are usually capable of being held open by a plain pull-up spindle with a cotter through it which rests on the bridge or gland of the valve. The cotter hole through the spindle is so situated that the cotter can only be inserted when the valve is open, otherwise the cotter hole is blind. In case of emergency, any valve which is not easily and readily accessible must be fitted with an extension device to the spindle to enable it to be operated quickly. For quick closing, some valves may be specially designed to open against the compression of a spring, the spindle passing through a collapsible bridge can be operated by a lever connected to it by rod or wire.

The usual shipside fittings include injection and overboard discharge for circulation of a steam condenser; suction and discharge for lubricating oil coolers, refrigerator and distiller; ballast tank filling and discharge; bilge discharge; deck-hosing and fire-fighting suctions; evaporator inlet and blow-down; sanitary supply and discharge from showers, baths, toilets, and wash-basins.

Other openings in the hull under engine-room control are stern tubes and fin stabilisers which are described elsewhere.

OILY-WATER SEPARATOR

Recent IMO regulations have limited the oil in water content discharged into permissible waters to 15 parts per million. Older generations of oily-water separator had problems at maximum throughput rate to comply with the 100 ppm of oil in the effluent regulation.

In order to comply with the 15 ppm, older separators have either to be replaced or updated by the addition of separating cartridges in the overboard water discharge line.

Fig. 53 shows diagrammatically an automatic oil-water separator which, it is claimed, can reduce the effluent to 2 ppm of oil in the mixture or less. In order to achieve this low level the separator incorporates concentric cylindrical oil coalescing cartridges through which the oily-water mixture is drawn by a positive displacement pump. The coalesced oil rises to the top of the separator where its accumulation is detected by an oil-water interface probe. When in the normal mode a controller is constantly monitoring the oil-water interface level and the overboard discharge. In the event of the effluent exceeding set limit the process is stopped and alarm given.

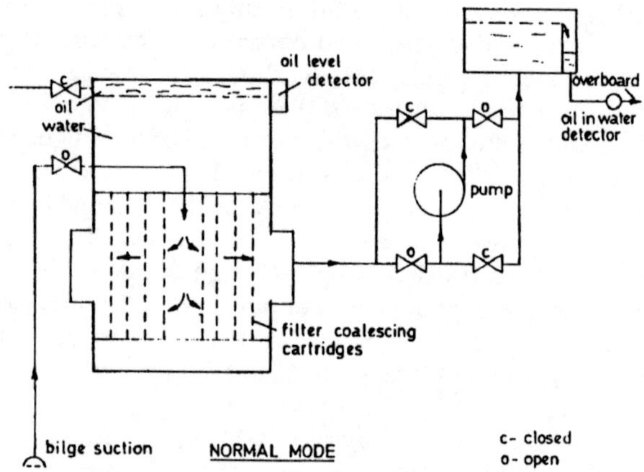

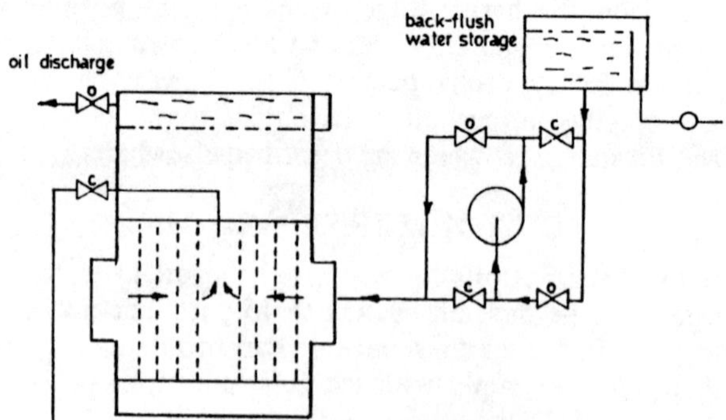

Figure 53
Cleaning mode
automatic oily-water separator

When the oil-water interface reaches its lower level the controller changes the operation to one of cleaning by back flushing and oil discharge. The oil-water interface will then rise to the higher level when reversion to normal mode takes place.

By using an oily-water separator in the suction mode rather than the delivery mode (*i.e.* the pump after the separator not before) disintegration of the oily-water mixture prior to separation is achieved, thus improving separation efficiency.

Oil-water interface probes are used to detect the level of the oil in the separator and to relay this detection to a controller and in some cases to give alarm. One such probe unit consists of two sheathed probes, one at a high level the other at a low level, the oil water interface somewhere between. The probes are connected to an a.c. circuit similar to the d.c. Wheatstone Bridge principle. Two coils are energised from the supply and two capacitor circuits complete the bridge, one capacitor connected to the probe being the variable in the circuit. The probe and tank form two electrodes of the variable capacitor whose capacitance depends upon the dielectric constant of the medium between. Thus the value of this capacitance depends upon the position of the oil-water interface. The bridge when balanced in air would become unbalanced in oil or water and the electrical signal change could be amplified and relayed. Similarly balance in oil would react to water, etc.

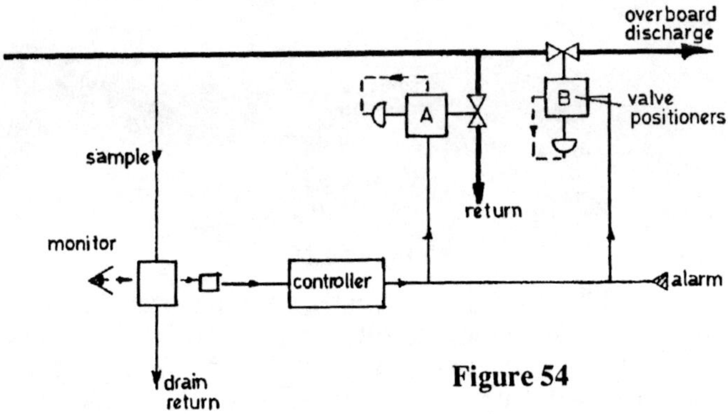

Figure 54

OIL IN WATER SENSOR.Using photo-cells oil in water can be detected. Fluid passing through a glass is exposed to long wavelength light from an U.V. lamp which causes fluorescence if oil particles are present. This light can be detected by the secondary element photo cell which produces a signal which is then amplified and can sound alarm or actuate a controller. The amount of fluorescence depends upon the amount of oil and this effects the visible light detected by the photo-cell. Use is made of the above oil in water sensor to monitor and control systems for discharges of oil-water mixtures. In Fig. 54 if the oil in water content is above the prescribed limit as set at the controller, valve A is automatically opened and B closed diverting the mixture and returning it to source.

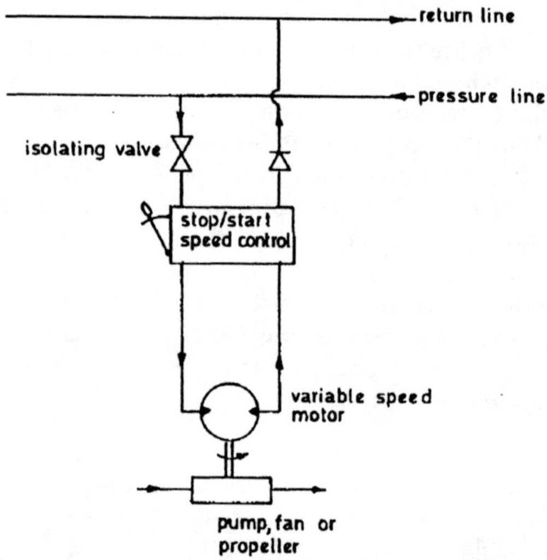

Figure 55

HYDRAULICS

A centralised hydraulic system consisting of duplicated oil pumps, accumulators, filters and an oil reservoir, fitted with pressure regulators which govern the pressure in different lines for different purposes is an economic, reliable and safe power installation.

Items that can be operated by such a system include: pumps, bow thruster, winches, windlasses, hatch covers, cranes, ramps, watertight doors, etc. The main advantages of a centralised hydraulic system are: 1. Smooth operation. 2. Infinitely variable speed control. 3. Intrinsically safe therefore useful for hazardous cargoes. 4. Centralised.

Fig. 55 shows how a controlled unit, pump, fan, etc, would be connected into the return and pressure lines which are connected to the centralised hydraulic system. The stop/start speed control may be fitted with a pilot line for remote operation.

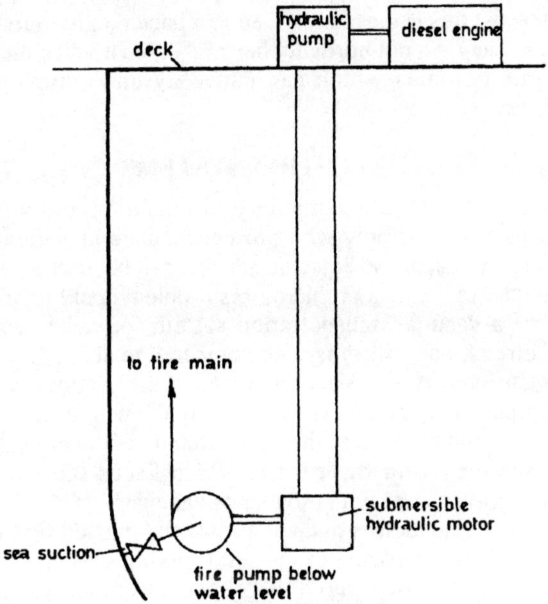

Figure 56

EMERGENCY FIRE PUMP SYSTEM

Fig. 56 shows a completely independent emergency fire pump system. The diesel would have its own source of fuel supply and be started in the same way as an emergency generator. Fire pump and hydraulic motor would be completely submersible and irrespective of the height and draught of the vessel the centrifugal fire pump would not require a priming device as it would be situated below the water line. Such an arrangement may also be used as a booster/priming device for a main fire pump situated on deck.

SEWAGE AND SLUDGE

Present regulations relating to sewage in the UK are simply that no sewage must be discharged whilst the vessel is in port. Standards adopted by other countries vary, but most are likely to conform to IMO recommendations, their main restriction is that the coliform count in the effluent discharged in restricted waters should not exceed 1000/100ml.

COLIFORMS: this is the name given to a bacteria group found in the intestines. They are not normally harmful, except when they contain pathogenic colonies which can cause dysentry, typhoid, para-typhoid, etc.

RETENTION SYSTEMS

Their main advantage is simplicity in operation and virtually no maintenance, they comply with present regulations within the limit of their storage capacity. Since no sewage can be discharged in port, prolonged stays create a problem, this problem could be reduced by the use of a vacuum transportation system for toilets where only about 1 litre of water/flush is used compared to about 12 litres/flush for conventional types. Vacuum systems use smooth, small bore plastic pipes (except in fire hazard areas) which are relatively inexpensive, and because of the small amount of water used they are usually supplied with fresh water which keeps salt water out of accommodation spaces with obvious advantages.

With some retention systems the sewage is passed first through a comminutor, which macerates the solids giving greater surface area. The mix is then passed into a chlorine contact tank where it must remain for at least 20 minutes before discharge overboard.

BIOLOGICAL TREATMENT PLANTS

PLANT DESCRIPTION: raw sewage passes through a comminutor into the collection compartment. When the level in this compartment rises sufficiently, overflow of the liquid takes place into the treatment-aeration compartment where the sewage is broken down by aerobic activation. Fluid in this compartment is continuously agitated by air which keeps the bacteriologically active sludge in suspension and supplies the necessary oxygen for purification.

The effluent is then pumped to a settlement compartment where the sludge settles out leaving treated effluent, which passes over a weir into the final compartment for chlorination before discharge overboard.

The settled sludge is continuously returned to the aeration compartment by an airlift pump.

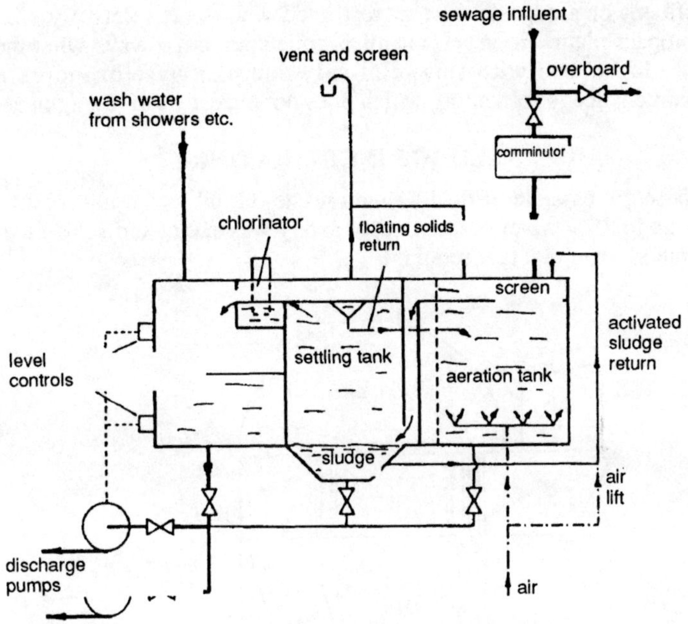

Figure 57
Extended aeration sewage plant

Excess sludge builds up in the settlement chamber and this must be discharged at regular intervals, in port this may not be possible hence it should be pumped to a sullage tank for disposal later. By using an incinerator to deal with excess sludge a sullage tank may not be required.

With the extended aeration system it can be 5 to 14 days before the plant is fully operational because of the prolonged aeration of sewage necessary to produce the bacteria that carry out the purifying process. Hence, the plant should be kept operational at all times. It should be noted that oil or grease entering the system kills useful bacteria.

CHEMICAL TREATMENT PLANTS

These are recirculation systems in which the sewage is macerated, chemically treated, then allowed to settle. The clear, sterilised, filtered liquid is returned to the sanitary system for further use and the solids are periodically discharged to a sullage tank or incinerator. The main advantages are (1) no necessity to discharge effluent or sludge in port or restricted waters (2) relatively small compact plant. However, chemical toilets are not always what they should be and with this relatively complex system increased maintenance is something which does not endear itself to engineers.

SLUDGE INCINERATORS

These are capable of dealing with waste oil, oil and water mixtures of up to 25% water content, rags, galley waste, etc, and solid matter from sewage plants if required.

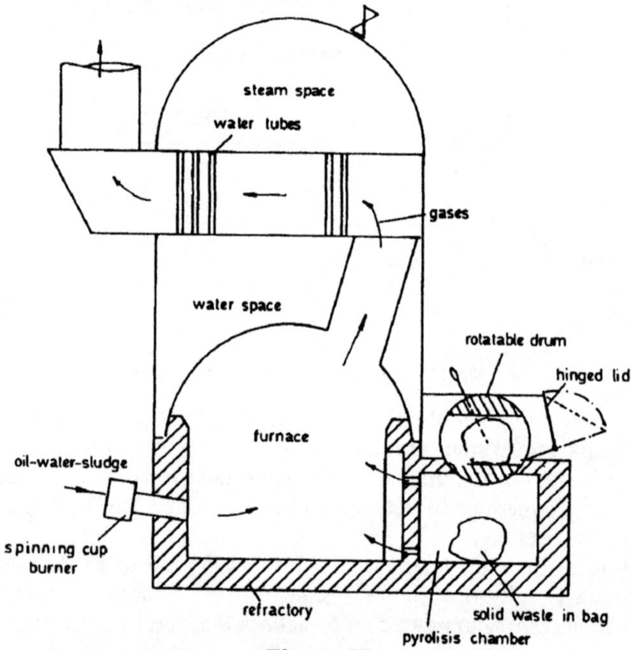

Figure 58
Incinerator

Fig. 58 shows a small water tube type of boiler combined with incinerator plant in order to provide an economy.

Homogenous oil/water mixtures that have been formed by passing them through a comminutor – a kind of grinder, macerator, mixer which produces a fine well dispersed emulsion – are supplied to the rotating cup burner. Solid waste from the galley and accommodation, etc, would be collected in bags and placed in the chamber adjacent to the main combustion chamber, the loading system of which is self evident in the diagram. The loading arrangement incorporates a locking device which prevents the doors (loading and ash pit) being opened with the burner on. The solid waste goes through a process that may be described as pyrolysis, that is the application of heat. Hydrocarbon gases are formed, due to the low air supply to this compartment, which pass into the main chamber through a series of small holes and burn in the furnace. Dry ash remaining in the chamber has to be removed periodically through the ash pit door.

Solid matter from sewage systems could be incinerated in this unit, a connection would have to be made from the sewage plant to the pyrolysis chamber of the incinerator.

CHAPTER SEVEN

REFRIGERATION

The natural transfer of heat is from a hot body to a colder body, the function of a refrigeration plant is to act as a heat pump and reverse this process so that rooms can be maintained at low temperatures for the preservation of foodstuffs.

Refrigerating machines can be divided into two classes, (i) those which require a supply of mechanical work which is the *vapour compression system*, and (ii) those which require a heat supply and work on the *absorption* system. The former is more efficient and in general use on board ship, the latter is not suitable for on board ship because the system includes vessels wherein correct and steady levels of liquid are critical for its correct working, therefore only the vapour compression system will be described here.

The refrigerating agent used in the circuit is a substance which will evaporate at low temperatures. The boiling and condensation points of a liquid depend upon the pressure exerted upon it, for example, if water is under atmospheric pressure it will vaporise at 100°C, if the pressure is 7 bar the water will not change into steam until its temperature is 165°C, at 14 bar the boiling point is 195°C and so on. The refrigerant used must vaporise at very low temperatures. The boiling point of carbon dioxide at atmospheric pressure is about −78°C (note *minus* 78), by increasing the pressure the temperature at which liquid $C0_2$ will vaporise (or CO_2 vapour will condense) is raised accordingly so that any desired vaporisation and condensation temperature can be attained, within certain limits, by subjecting it to the appropriate pressure.

Desirable properties of refrigerants:

1. Low boiling point, otherwise operation at high vacua becomes necessary.

2. Low condensing pressure.
3. High specific enthalpy of vaporisation.
4. High critical temperature.
5. Non corrosive.
6. Stable under all working conditions.
7. Non-flammable and non-explosive.
8. Non toxic.

Some of the agents employed as refrigerants and their more important characteristics are as follows:

Carbon dioxide ($C0_2$) This is an inert vapour, non-poisonous, odourless, and has no corrosive action on the metals. Its natural boiling point is very low which means that it must be run at very high pressures to bring it to the conditions where it will vaporise and condense at the normal temperatures of a refrigerating machine. A further disadvantage is that its critical temperature is about 31°C which falls within the range of sea water temperatures. At its critical temperature and above, it is impossible to liquefy the vapour no matter to what pressure it is subjected and, as part of the circuit depends upon condensing the vapour in a condenser circulated by sea water, great difficulty is experienced, even with special additional devices, when in high temperature tropical waters.

For these reasons carbon dioxide vapour compression installations are only to be found on older vessels. However carbon dioxide can be used in containers, it is injected into them in liquid form to produce low temperatures – "snow shooting".

Ammonia (NH_3). This is a poisonous vapour and therefore an ammonia machine should not be open to the engine room but have a compartment of its own so that it can be sealed off in the case of a serious leakage; water will absorb ammonia and therefore a water spray is a good combatant against a leakage. Ammonia will corrode copper and copper alloys and therefore parts in contact with it should be made of such metals as nickel steel and monel metal. Its natural boiling point is about –39°C therefore the pressures required throughout the system to obtain the necessary evaporation and liquefaction temperatures are much lower than that required in a $C0_2$ machine. Ammonia is a superior refrigerant thermodynamically than carbon dioxide but its disadvantages, *e.g.* will taint foodstuffs, will form explosive mixtures with air, in addition to those already outlined restrict and reduce its use.

Purpose made primary refrigerants R12, R22 and R502 (the R stands for refrigerant, the digits are part of a code system for the number of atoms making up the molecule in these fluorinated hydrocarbons) are the main refrigerants in use today. Their properties are similar but R22 and R502 are used for lower temperatures than the R 12.

R. group of refrigerants have most of the desirable properties listed before with the additions and exceptions of the following:

1. Miscible with oil *i.e.* they remove oil films so separators may be required.
2. Toxic fumes are given off in the event of fire – they decompose to liberate phosgene gas.
3. Low specific enthalpy of evaporation, compared to ammonia about one third.
4. It is believed that they damage the ozone layer.

All of the foregoing refrigerants are called primary as they are used in the first circuit of the system in relatively small amounts. Secondary refrigerants are *e.g.* Brine or Air (a combination of all three gives primary, secondary and tertiary), they are used in large quantity for practical simplicity and economic reasons.Imagine having a direct expansion refrigeration system in a hold compartment with vast quantities of expensive R22 circulating – you would require a large compartment to store a recharge of R22 in the event of leakage.

WORKING CYCLE

Fig. 59 illustrates diagrammatically the essential components of the vapour compression system, which consists of the *Compressor* driven by an electric motor, the *Condenser* which is circulated by sea water, the *Regulator* (sometimes referred to as the expansion valve), and the *Evaporator* which is circulated by brine. It is a completely enclosed circuit, the same quantity of refrigerant passes continually through the system and it only requires to be charged when there are losses due to leakage. Any of the above agents may be used as the refrigerant, the difference in their working being only the pressures throughout the system, therefore in the following description the words "refrigerant," "the liquid" and "the vapour," etc., will be used.

The refrigerant is drawn as a vapour at low pressure from the evaporator, into the compressor where it is compressed to a high pressure and delivered into the coils of the condenser in the state of a superheated vapour. As it passes through the condenser coils, the vapour is cooled and condensed into a liquid at approximately sea water temperature, the heat given up being absorbed by the sea water surrounding the coils and pumped overboard. The liquid, still at a high pressure, passes along to the regulator, this is a valve just partially open to limit the flow through it. The pipeline from the discharge side of the compressor to the regulating valve is under high pressure but from the regulating valve to the suction side of the compressor is at low pressure due to the regulating valve being just a little way open. As the liquid passes through the regulator from a region of high pressure to a region of low pressure, some of the liquid automatically flashes off into a vapour absorbing the required amount of heat to do so from the remainder of the liquid and causing it to fall to a low temperature, the temperature being regulated by the pressure, so that the refrigerant enters the evaporator at a temperature lower than that of the brine. The liquid (more correctly a mixture of liquid and vapour) now passes through the coils of the evaporator where it receives heat to evaporate it before being taken into the compressor to go through the cycle again. The required heat absorbed by the liquid refrigerant in the coils of the evaporator to evaporate it is extracted from the brine surrounding the coils which causes the brine to be cooled to a low temperature.

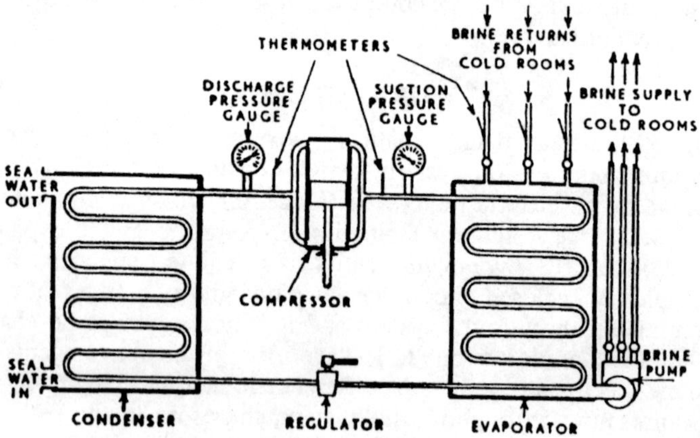

Figure 59
Vapour compression refrigerating machine

BRINE CIRCULATION

The cold brine from the evaporator is pumped through pipes led around the top of the walls and sometimes also the ceilings of the cold storage rooms, it extracts heat from the rooms to cool or maintain them at a low temperature, and returns slightly warmer in consequence of absorbing heat, to the evaporator.

The simplest form of brine cooling is the *single-temperature brine system* wherein there is only one evaporator in the refrigerator to produce brine at one temperature only, and the same temperature brine is pumped through the cooling pipes of the various cold rooms which require to be kept at different temperatures. Provided that the temperature of the brine is below that of the coldest room, then the *quantity* of brine flowing through the pipes of any one room determines the temperature of that room. Therefore different rooms can be maintained at different temperatures by controlling the quantity of brine flowing through their separate pipe circuits. This is done by regulating the amount of opening of the brine return valves at the evaporator, the brine pump discharge valves are left full open to avoid air-locks in the circuits and only closed when repairs are required.

In addition to the room thermometers there is also a thermometer at each brine return, the temperature of the warm brine returning from any particular room provides a good indication of the temperature within that room.

When there is considerable variation of temperature conditions in ships carrying large quantities of food, such as frozen meat in one room at −8°C, some kinds of soft green fruit in another at over 10°C, with other rooms at varying temperatures between these, it is usually the practice to have two or three separate brine circuits maintained at different temperatures by separate evaporators in the refrigerating plant. These are referred to as *multi-temperature brine systems*.

In fruit carrying rooms the air must be continually circulated to prevent stagnant pockets of CO_2 forming around the fruit, therefore the cold air system of cooling is employed. Instead of wrapping the brine pipes around the walls and ceiling inside the rooms, the brine is led into grid-boxes through which the air, drawn from the bottom of the rooms by fans or blowers, is passed over the brine grids and blown back into the rooms via ducting along the ceiling, see Fig. 60.

The brine is made by dissolving calcium chloride in fresh water, the freezing point depending upon the density, but the liquid should not be too thick as to impede its free flow through the pipes. A satisfactory mixture is obtained by 0.25kg of calcium chloride per litre of fresh water. This will produce a brine with a freezing point well below any working temperature.

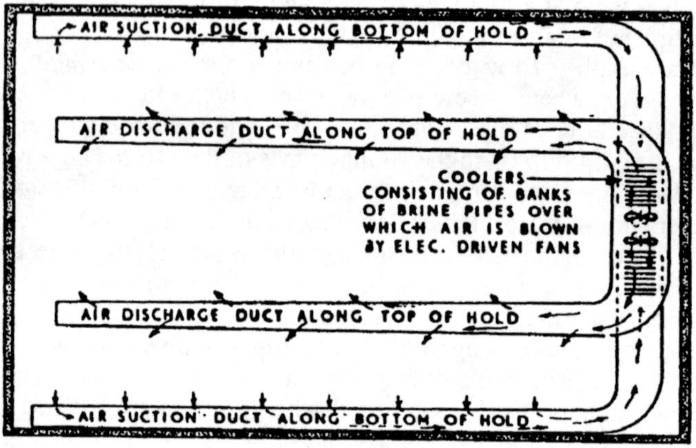

Figure 60
Cold air system of cooling

In the event of an abnormal leakage of brine with consequent running short of calcium chloride, an emergency brine may be made with sodium chloride (salt) and fresh water, but an alkali (such as caustic soda) should be added to render the brine slightly alkaline (test by litmus paper) and prevent corrosion. On no account must sea water be used to replace brine, its freezing point is not sufficiently low for the average working temperatures of brine and it is highly corrosive.

REFRIGERATED CARGO VESSELS (REEFERS)

These are mainly containerised vessels and containers carried fall principally into two types: porthole containers connected to air ducts into which cooled air from a central refrigeration plant is supplied, Fig. 61 shows the arrangement, and independent containers with an integral cooling system. There are variants between the two.

Refrigeration machinery used in a central refrigeration plant is usually R22 multi-cylinder reciprocating compressors with up to 16 cylinders, belt driven or direct drive with speeds up to 1750 rev/min. Screw types of compressor are becoming increasingly popular because of their greater reliability and reduction in maintenance. Multiple inter-connected units, where a unit consists of compressor; condenser and sub divided cooler or separate coolers in each chamber, are required by classification societies.

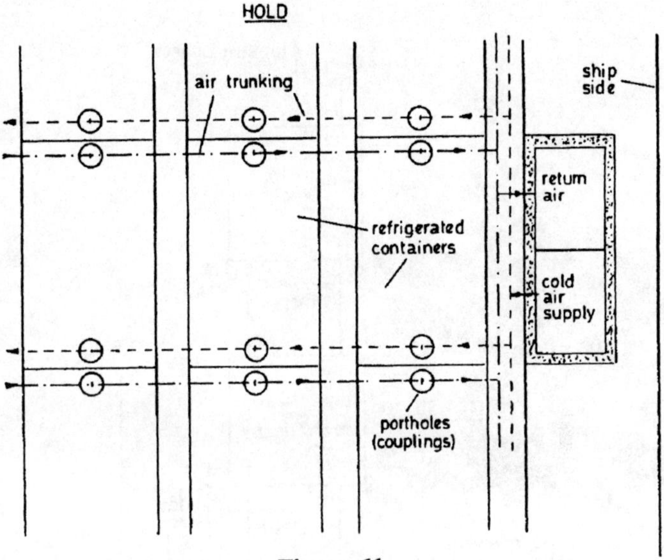

Figure 61

Stand-by equipment is also provided, so that if at maximum load two compressors are sufficient, three are provided. Brine circulation to air coolers as a secondary refrigerant enables, by means of pumps valves and automatic controlled circulation of brine, holds to be maintained at different temperatures.

Refrigerated cargo requiring fresh air changes to reduce carbon dioxide and ethylene concentration may have, in new vessels, heat recovery economy installations fitted. These may be recuperative or regenerative, the former being the more popular. Fig. 62 shows diagrammatically the two types.

Recuperative: cross stream plate heat exchangers, compact and space saving. Cooling the fresh air down to exhausted air temperature before it goes to the main cooler and giving an energy saving of about 38%.

Regenerative: a slowly rotating cylindrical heat exchanger filled with corrugated aluminium strip. Part of the heat exchanger in the outgoing air cools down, the other part cools the incoming air and heats up giving an energy saving of about 45%.

The energy saving depends upon: 1. Type of heat exchanger system used; 2. Product temperature required; 3. Frequency of air changes; 4. Outside air humidity and temperature.

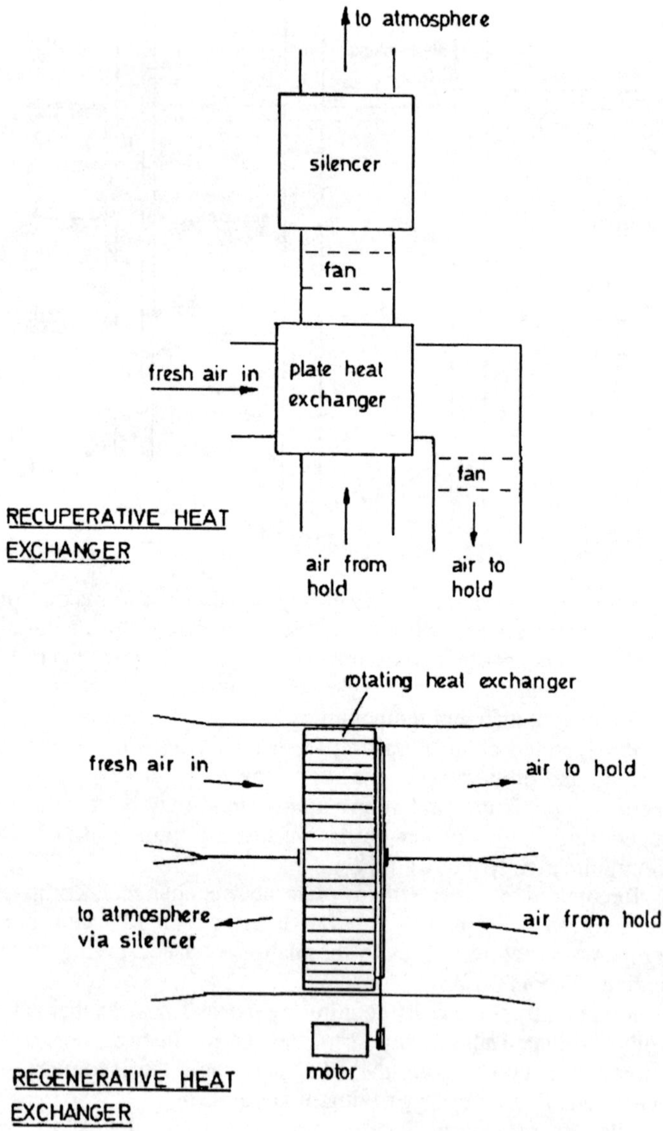
to atmosphere
silencer
fan
fresh air in
plate heat exchanger
fan
RECUPERATIVE HEAT EXCHANGER
air from hold
air to hold
rotating heat exchanger
fresh air in
air to hold
to atmosphere via silencer
air from hold
motor
REGENERATIVE HEAT EXCHANGER

Figure 62

DIRECT EXPANSION SYSTEM

In small refrigerators for ships' provisions, the evaporator with brine circulation as previously described can be dispensed with and the refrigerant, after passing through the regulator, is evaporated and expanded directly in the grid coils on the ceiling and walls of the cold room. This is known as the *direct expansion system.* The condenser may be water or air cooled, depending upon its loading, in very small and domestic types air cooling is usually sufficient. R12 is most often used as the refrigerant and it works on the vapour compression cycle of operations like any other compression refrigerator.

The plant is fully automatic and has a thermostatic control switch to start and stop the machine, usually over a temperature differential setting of about 2°C. This means that if the cold chamber is to store vegetables at 0°C, the control switch would be set to cut out at –1 °C and start up at +1°C.

INSULATION

All cold rooms must be insulated against heat flowing into them from the outside, this is done by building a wood wall around the room leaving a space between it and the steel plating, and filling up this space with some light-weight heat insulating material such as cork or silicate of cotton. The inside of the wood wall is usually protected against damp by sheet zinc. In some cases a double wall is built to allow an air space between the shell and the insulated wall (see Fig. 63).

The hatch covers are also insulated. One method is by constructing them in the form of hollow rectangular boxes or plugs filled with cork. The sides are inclined so that one interlocks another, necessitating the removal of the end plugs (keys) before lifting the others.

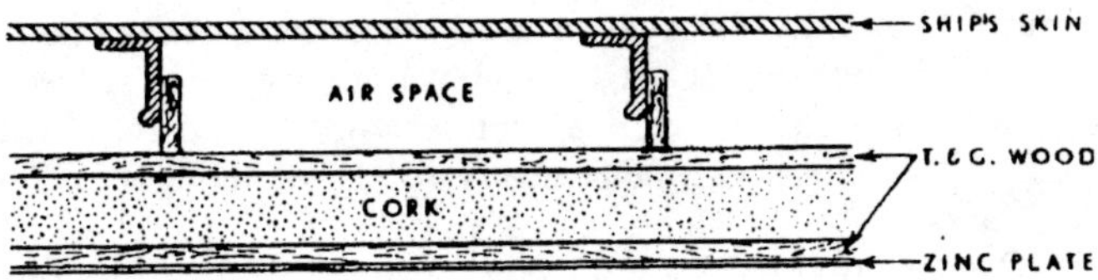

Figure 63
Insulation of cold chambers

TEMPERATURES OF FOOD STORAGE

The temperatures at which the various foodstuffs are stored may vary considerably depending upon varying conditions such as time in storage, whether fruit is taken on board green and is to be almost ripe at time of discharge, and so on. Some average storage temperatures are given below:

	°C		°C
Lemons (green)	13	Vegetables	1
Figs and dates	12	Eggs	0
Bananas (green)	11	Apples and pears	0
Tomatoes	10	Chilled meat	−1
Wines and bottled beer	6	Furs and fabrics	−3
Tobacco	5	Fish	−7
Oranges	4	Poultry	−8
Grapefruit	2	Frozen meat	−8
Milk	1	Butter	−10
Cheese	1	Frozen fish	−15

LIQUEFIED GAS CARRIERS

Various fluids which at ambient temperature and pressure would be gases can be transported in liquid form. Liquefied gas cargoes can be divided into three main groups:

1. Liquefied natural gas (LNG)
2. Liquefied petroleum gas (LPG)
3. Chemical

In order to maintain the cargoes in liquid form various systems are used, system choice depends upon cargo properties, voyage duration, economics and other factors.

LNG. Mainly methane (CH_4) mixed with other hydrocarbon gases and possibly nitrogen.

Properties: boiling point of methane is about −160°C at atmospheric pressure and its critical temperature is −83°C at 47 bar pressure. Generally the system used keeps cargo temperature at −165°C and pressure very slightly above atmospheric. To keep the cargo at just below its critical temperature would require high pressures, which would mean heavy robust containers and some refrigeration.

IMO code for LNG carriers states that cargo pressure must be kept below the maximum relief valve setting by one or more of the following: 1. Refrigeration of the cargo; 2. Use of boil-off as a fuel in the propulsion system or vent it to atmosphere. Boil-off rate varies between 0.15 to 0.25%/day on voyage, some of the boil-off vapour is generally used for the main propulsion machinery, the rest is normally vented to atmosphere but it could be re-liquefied. Cost of re-liquefaction can be high.

LPG. Hydrocarbons such as propane, butane, butadiene, etc. Properties: boiling point and critical temperatures from below to above ambient depending upon the cargo, *e.g.* propane critical temperature 97°C, butadiene 152°C. The system used depends upon the cargo and may be: 1. Fully pressurised without a refrigeration plant; 2. Semi-pressurised with a refrigeration plant for re-liquefaction either in part or in total; 3. Fully refrigerated.

CHEMICAL. Ammonia, vinyl chloride, ethylene are examples and the properties are wide and various hence the system used could be as for LPG or LNG. Carriers often have multi-purpose capability.

Fig. 64 shows a system for re-liquefaction of boil-off. Helium or nitrogen refrigerants may be used for low temperature cargoes. Nitrogen can be obtained from LNG boil-off.

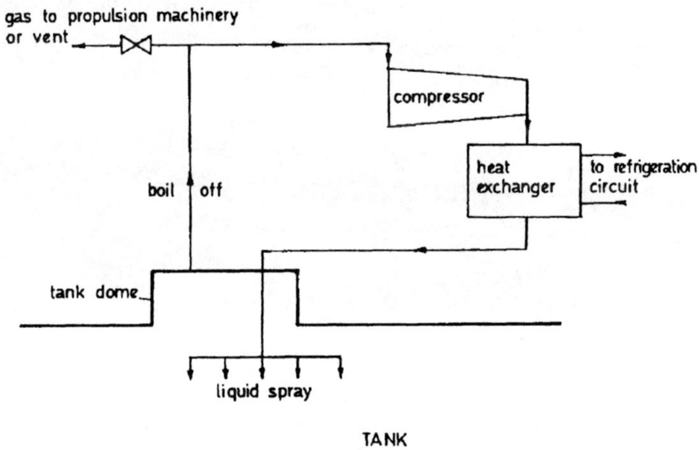

Figure 64

CHAPTER EIGHT

STEERING GEARS

There are three main types of steering gears employed for power operation of the rudder and are classified with regard to the type of unit which provides the power, (i) the steam steering gear where a steam engine provides the power, (ii) the hydraulic steering gear where the power is applied by rams working in hydraulic cylinders, (iii) the electric steering gear where the power is provided by electric motors.

STEAM STEERING GEAR. This type is now becoming obsolete but there are many of them still in service. It consists of a two-cylinder steam reciprocating engine with the respective cranks of each piston at 90 degrees to each other. This enables the engine to start from any position, if one crank and its piston is on the top or bottom centre, the other crank will be at right angles and its piston at about half-stroke and ready to start as soon as the steam control valve is opened. Each cylinder has its own piston valve with no lap or lead so that steam is supplied to the cylinder for the full stroke, economy is thus sacrificed for reliability of immediate starting and running.

TELEMOTOR CONTROL

The telemotor system is a remote control of the steering engine control lever from the bridge wheel by hydraulic transmission. The bridge steering wheel is geared to a plunger inside a hydraulic cylinder, this is the transmitting cylinder and is connected at its two ends to the two ends of a receiving cylinder near the steering engine by two small diameter heavy gauge solid drawn copper pipes. The receiving cylinder contains a plunger and rod with a crosshead connection to the steering engine control lever.

The circuit is filled with a non-freezing lubricating liquid such as a mineral oil. The liquid, being non-compressible, causes the plunger in the receiving cylinder to follow slavishly any movement of the plunger in the transmitting cylinder. Springs are fitted to the moving parts at the receiver cylinder to assist the gear to return to its mid position. The plunger rod at the receiving cylinder is connected to the steering engine control lever, thus when the helmsman moves the steering wheel on the bridge there is an immediate movement of the control valve to start the steering engine.

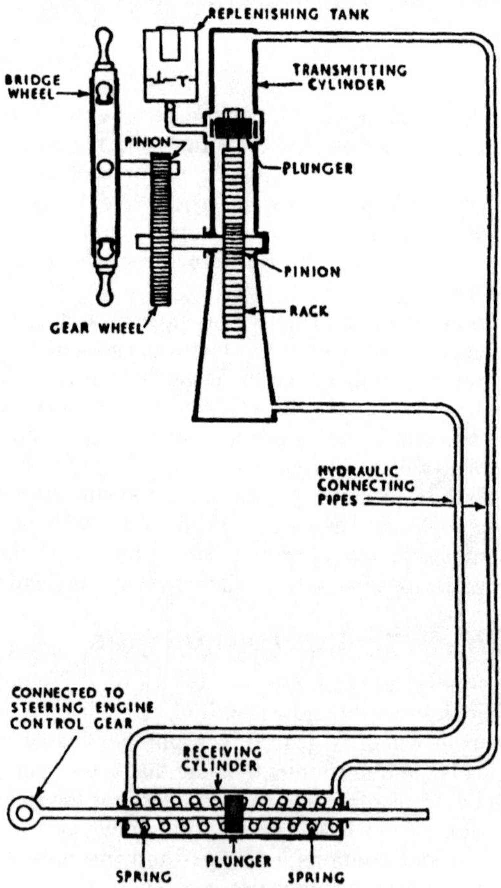

Figure 65
Principle of the telemotor

Telemotor systems vary considerably in design, twin cylinders are often used instead of a single cylinder; the receiving cylinder with a centre division plate is often the moving unit to actuate the steering engine control lever, with stationary hollow rams protruding into each end. The return springs are usually situated outside the cylinder. A charging system is incorporated to pump through the cylinders and pipe lines to expel any air bubbles and ensure a solid rail of liquid throughout the circuit.

Fig. 65 therefore is a very much simplified diagram without valves and charging system, to illustrate the principle of the telemotor.

A CREEP TEST of the telemotor gear is usually carried out before the ship is ready to leave port. This consists of putting the steering wheel hard over in one direction and lashing the wheel to hold it there. The position of the receiver crosshead connection to the steering engine control is marked and the system is inspected to see that there are no leakages. After a reasonable time (say half-an-hour), the position of the receiver crosshead is checked to see if any slip or creep has occurred. A similar procedure is carried out with the wheel lashed hard over in the opposite direction. If there are no leakages and no slip from the hard over to port or starboard positions, it indicates the telemotor system to be tight and in good condition.

CHARGING. Air in the telemotor system is dangerous and must be avoided, it is compressible and therefore will cause a time lag between steering wheel movement on the bridge and steering engine response. Large quantities of air will cause sluggish and faulty steering.

Fig. 66 is a diagrammatic sketch of a charging system. The non-return valve on the return pipe to the charging tank is spring-loaded, its function is to prevent back-flow of oil due to gravity from the highest point, and air entry at the open end, when the charging valves are open.

To charge means to pump through and rid the system of all air. The charging tank is filled (a full head must be maintained all the time during charging), steering wheel put and held in mid position, by-pass valve opened, charging valves opened, and the hand-pump operated until the oil returns to the charging tank in spurts corresponding immediately to the delivery strokes of the pump, thus indicating that there is no air in the system to cushion the oil flow. During pumping through, the air cocks may be opened to get rid of any excess air and closed when oil escapes.

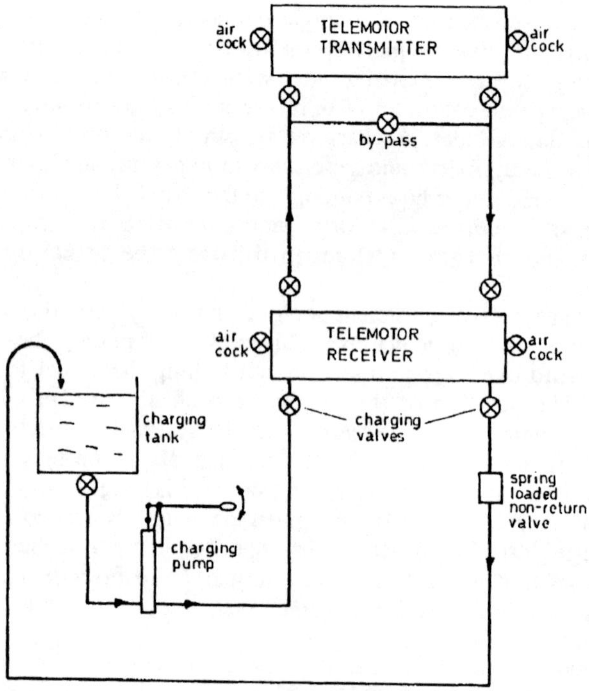

Figure 66
Telemotor charging

RUDDER STOCK UNIT

The steering engine or electric motor transmits its movement to the tiller firmly keyed to the rudder stock, by means of a worm on the engine crank shaft which engages with a worm wheel, the shaft of this worm wheel carries a pinion which meshes with a large quadrant the centre of which sits loosely over the head of the rudder stock above the tiller. Two heavy shock absorbing helical buffer springs connect the two sides of the fixed tiller to the loose quadrant. See Fig. 67 and note that it shows diagrammatically that when the quadrant is moved it pulls the tiller with it through one of the springs which takes the load in *compression*. The function of the springs is to absorb the shock of heavy seas striking the rudder and so prevent damage to the steering engine and the teeth on the driving pinion and quadrant.

An emergency hand steering gear may be fitted to drive a pinion engaging with a toothed quadrant extension secured to an arm on the tiller, the drive being from a handwheel carried on a pedestal above the steering gear, through worm gearing, friction clutch and vertical shaft down to the quadrant driving pinion.

The shackle shown on the lug of each wing of the loose quadrant is for coupling block and tackle gear to operate the rudder in the event of breakdown of the steering engine (and also the emergency hand steering gear if one is fitted). The block and tackle arrangement is worked through wire ropes, guided by pulleys, and led to the after winch.

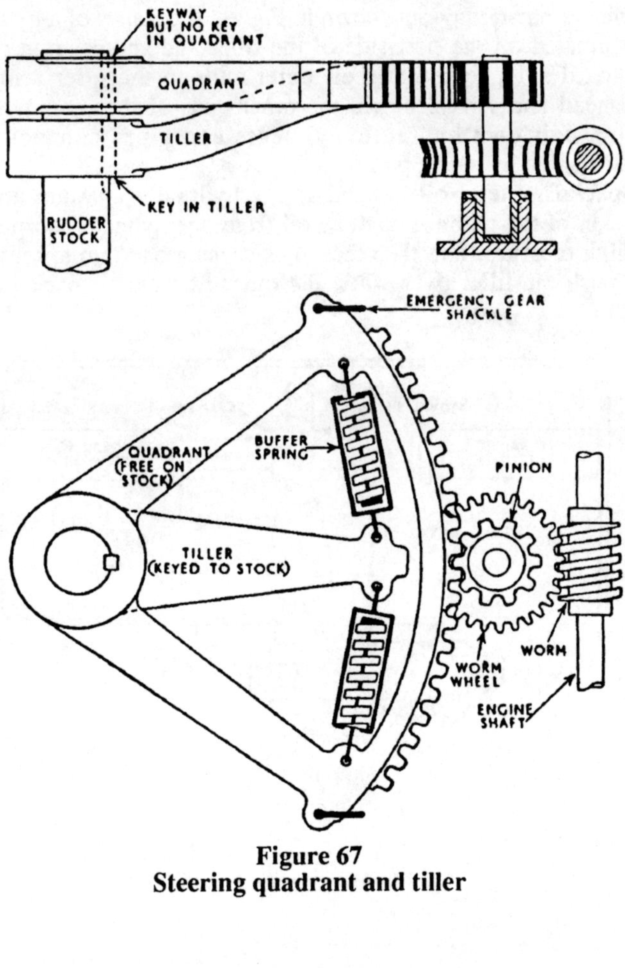

Figure 67
Steering quadrant and tiller

A screw-operated brake is fitted to enable the rudder stock to be locked while changing over from engine to emergency steering, or while repairs are being carried out, and the steering engine bedplate is mounted on a sliding base so that the engine and driving pinion can be slid out of gear after the emergency gear has been coupled up.

The foregoing system will only be encountered on older and small vessels,but its simplicity, reliability and cheapness are advantages.

ELECTRO-HYDRAULIC STEERING GEAR

This is perhaps the most popular type of steering gear. Referring to the diagrammatic lay-out shown in Fig. 68 it consists of a hydraulic ram situated on the port side of the tiller and another ram on the starboard side, linked at their outer ends to the tiller arm by a crosshead and swivel block, the other ends of the rams working inside their own hydraulic cylinders and pipes connect these cylinders to a hydraulic pump.

Special mineral oil is used as the hydraulic medium and the function of the pump is to draw oil from one cylinder and pump it (at high pressure) into the other, thus causing one ram to move out and push the tiller over while the other ram moves back into its

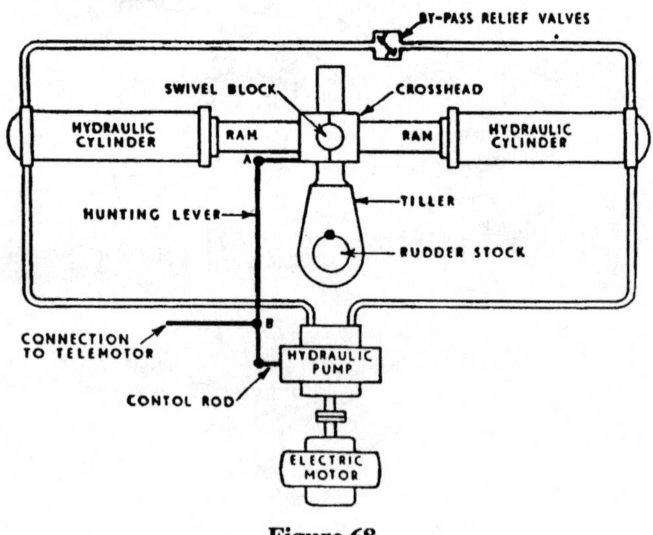

Figure 68
Hydraulic steering gear

cylinder. The hydraulic pump is of the rotary displacement type driven continuously by an electric motor (hence the name of this steering gear being *electro*-hydraulic, if a steam engine was the driving unit of the pump it would be termed a *steam*-hydraulic steering gear). The pump is of special construction and may be a Hele-Shaw or Williams-Janney design, it runs continuously in the same direction and the position of a movable plate inside the pump controls the suction and discharge of the oil. When the plate is in mid position, no oil is drawn in nor discharged, when the plate is moved in one direction from mid position oil is drawn from one cylinder and discharged into the other, when the plate is moved in the opposite direction the suction and discharge of the oil is reversed in direction. The plate is actuated by a control rod which is attached at its outer end to the hunting gear.

If a heavy sea strikes the rudder, the shock is transmitted through the tiller to the rams, this causes a sudden increase in oil pressure in one of the cylinders and double spring-loaded relief valves allow the tiller to give way slightly by by-passing a little of the oil into the other cylinder. The resultant displacement of the rudder, tiller and ram crosshead moves the pump control rod through the hunting gear and the tiller is automatically brought back to its proper position.

THE HUNTING GEAR for this type of steering gear is a simple arrangement of levers and will be readily understood by reference to Fig. 68. If the telemotor link is moved to the right, the hunting lever will swivel about A as a fulcrum and the pump control rod will be pushed inwards, the pump will then draw oil from the right cylinder and discharge it into the left and the crosshead (and tiller) will begin moving to the right. As the crosshead moves, B now acts as the fulcrum for the hunting lever and the movement of A to the right will cause the other end, connected to the control rod, to move outwards to bring the control plate in the pump back to its mid position, the pump will cease to deliver oil and the gear will come to rest.

A FOUR-RAM HYDRAULIC STEERING GEAR may be fitted in large ships for greater steering power, instead of the two-ram type described above. The four-ram unit is simply a double two-ram unit, the tiller having a double arm so that the force of two diagonally opposite rams can act on the tiller to produce double the turning effect. A line diagram of a four-ram hydraulic steering gear is shown in Fig. 69.

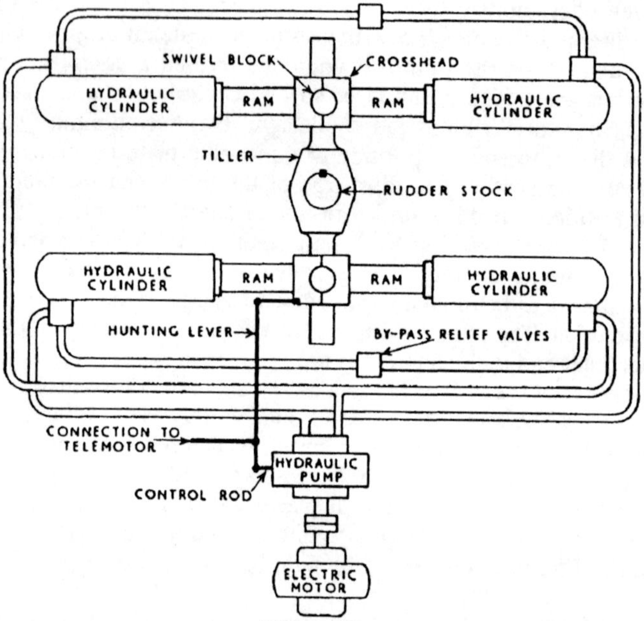

Figure 69
Four ram hydraulic steering gear

ROTARY VANE STEERING GEAR

The rotary vane type of steering system is a more recent development of the hydraulic steering gear and is shown in diagrammatic form in Fig. 70.

It consists of a rotor which is a taper fit on the rudder stock and keyed to it, and a stator of larger internal diameter than the outside diameter of the rotor, to form an annular space between them, the stator being firmly fixed to the ship's structure to prevent it rotating. The rotor has equidistantly spaced outwardly projecting radial vanes, and the stator has similar vanes projecting inwards, the spaces between the vanes form segmental pressure chambers for the high pressure hydraulic oil supplied from variable delivery pumps. Oil sealing between the ends of the vanes and their opposite working surface is effected by rubber-backed steel strips in grooves in the vanes.

The pressure chambers between the rotor and stator vanes are divided into two sets so that when the oil at a high pressure is supplied to one set and drawn from the other, the rotor will be forced to rotate in one direction turning the rudder stock with it, and by reversing the flow of oil the rotor (and rudder stock) will turn in the opposite direction. For simplicity, the action for one direction only is shown in the sketch. Allowing for the thickness of the vanes, a unit of three rotor vanes and three stator vanes will permit a rudder movement of 35 degrees to extreme port or starboard from mid position, *i.e.*, a total angle of 70 degrees, the vanes also acting as rudder stops. Relief valves and by-pass valve are incorporated in the oil system to absorb rudder shock.

Compared with the four-ram hydraulic steering gear, the cost of the rotary vane type to produce the same torque on the rudder stock is generally less, it is lighter in weight, takes up less space and requires less maintenance.

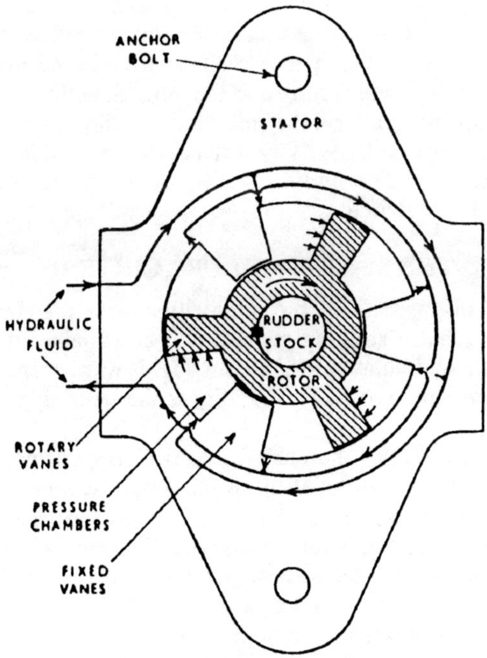

Figure 70
Rotary vane steering

ELECTRIC STEERING GEAR

In this type of steering gear the rudder stock is moved by the loose-quadrant and fixed tiller arrangement (Fig. 67) an electric motor suitably geared down to the quadrant drive.

This steering gear has the advantage over the others in that all connections from the bridge to the steering engine compartment consist of electric cables, no telemotor or mechanical control being needed.

The strength and direction of the magnetic field of the steering gear electric generator (driven by an electric motor off the mains) is controlled by a rheostat operated by the bridge wheel. When the wheel is turned, electric current from the mains is allowed to pass through the field coils of the steering gear generator to produce a magnetic field so that it generates electric current in a certain direction. This current is carried to the electric rudder motor and causes it to run and move the rudder. As the rudder motor turns, it operates another rheostat which electrically balances the bridge rheostat, this stops the current supply to the field coils of the generator and it ceases to generate, the rudder motor consequently comes to rest. Thus the rudder rheostat acts as the hunting gear. When the bridge wheel is turned in the other direction, current flows through the field coils of the generator in the reverse direction, current of reverse polarity is generated and the rudder motor runs the opposite way. The hunting action of the rudder rheostat will again bring the system to rest.

THE SINGLE FAILURE CRITERIA

The basic rule agreed by the authorities after the *Amoco Cadiz* disaster was that a steering gear should be capable of recovering from a single hydraulic failure within 45s. It was accepted that after a failure a minimum of 50% capability of the steering gear would be available.

Steering gears must therefore have to be duplicated, this would ensure 100% torque in the event of failure but would be expensive, or be capable of operating in two halves that can be completely isolated from one another and thus give 50% torque minimum in the event of a single failure. Rudder movement and operational speed would be reduced in the latter case.

A simplified diagram of a split system 4 ram gear is shown in Fig. 71. In the event of a single failure the following applies:

1. One of the two independent systems will be isolated and the other will continue to operate automatically.

2. Steering will be adequately maintained without significant repair.
3. Before serious oil loss automatic isolation of the leak occurs.
4. Alarm will be given.

If total steering failure takes place the rudder can be locked manually.

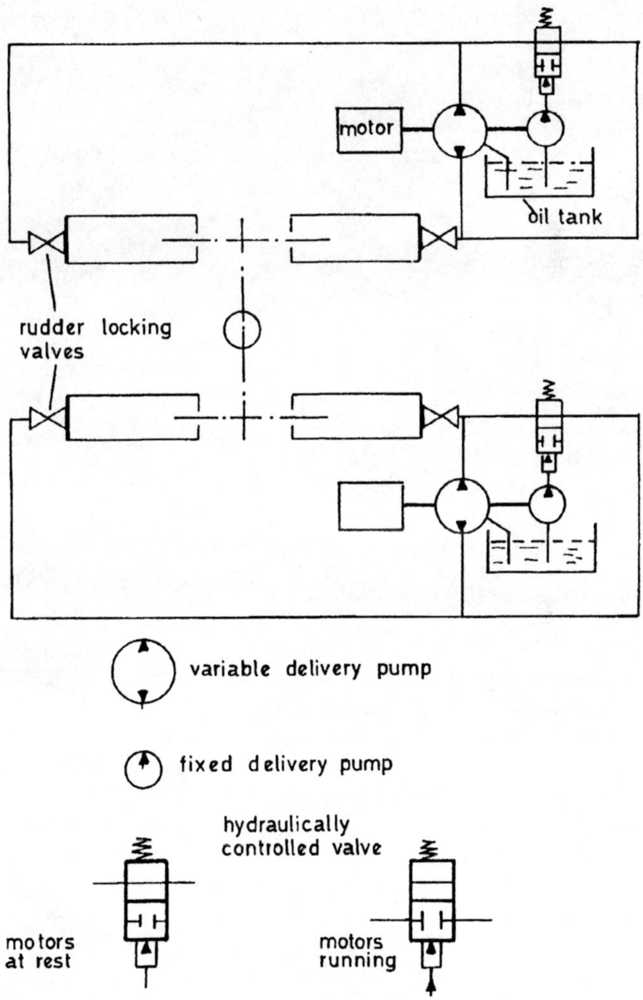

Figure 71

CHAPTER NINE

MAIN SHAFTING, PROPELLER, FUEL CONSUMPTION

The engine power is transmitted to the propeller by a line of forged steel shafting, comprising (i) thrust shaft, (ii) intermediate shafts, (iii) propeller shaft, often referred to as the tail-end shaft.

THRUST SHAFT AND THRUST BLOCK

The thrust shaft is connected to the main engine crank shaft in the case of a direct drive reciprocating engine, or to the main gear wheel shaft in geared installations. Its function, as well as transmitting the engine torque along to the next shaft, is to transfer the thrust of the propeller to the thrust block which, being securely fixed to the ship's structure, transmits the thrust to the hull of the ship.

The shaft is comparatively short with a coupling at each end, a thrust collar in the middle of its length and a journal at each side of the thrust collar. The journals run in bearings housed in the thrust block which carry the weight of the shaft.

Each side of the collar bears upon a number of kidney shaped white-metal-faced pads supported in the thrust block, those on the forward face of the collar being to take the ahead thrust, those on the after face to take the astern thrust. The back of each kidney piece has a hump or stem to allow the pads to pivot and tilt slightly so that the lubricating oil, picked up by the collar from the bottom of the block, can squeeze its way as a wedge-shaped film between the pad and collar surface, and be dragged over the whole surface. Thus there is always a film of oil maintained between the faces and there is consequently no metallic contact. Thrust pressures in the region of 24 bar can therefore be carried without danger of overheating due to friction. A single collar thrust block is illustrated in Fig. 72.

STRENGTHENING OF DOUBLE BOTTOMS. Particular attention is given to the strengthening of the structure of the double bottom in way of the thrust block since, not only must the heavy loads be supported and vibration minimised, but the thrust from the thrust block is to be transmitted to the ship's hull.

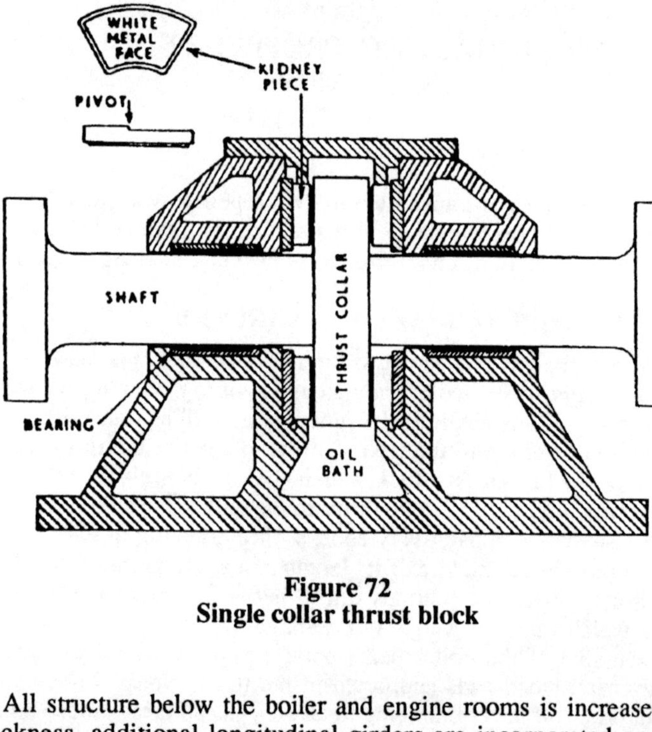

Figure 72
Single collar thrust block

All structure below the boiler and engine rooms is increased in thickness, additional longitudinal girders are incorporated so that they are pitched closer together, all girders have double angles, and all parts are a bearing fit. A tank top plate of extra thickness (40mm or more) runs continuously from under the engine bed plate to under the thrust block seating, the forward edge of the thrust block base either contacts the engine bedplate or chocks are fitted to have the same effect of spreading the thrust load over a greater area of the ship's hull. Chocks are also fitted at the after edge of the thrust block base. Most modern diesels have the thrust block as an appendage to or integral with the engine.

INTERMEDIATE SHAFTS AND PLUMMER BLOCKS

The intermediate shafts are those which connect the thrust shaft to the propeller shaft, the number of intermediate shafts depending upon the distance between the thrust block and the stern tube. Each shaft has a coupling at each end to connect by coupling bolts to the next, and has one, sometimes two, journals which run in white-metal lined bearings to carry the weight. The bearings are mounted on pedestals built up from the tank top.

A *plummer-block*, by definition, is a metal bearing for supporting a revolving shaft, with a removable cover to give access to the bearing. The bearings supporting the intermediate shafting come into this category and are therefore often referred to as plummer blocks. All these bearings with the exception of the aftermost are white metal lined on the bottom half only, which may be a plain surface with oil grooves or a set of pivoted (Michell) bearing pads. All the bearings are water-cooled.

The aftermost plummer block is white metal lined over the top half as well as the bottom. This is called the *trailing block* if the shaft which passes through it has a trailing collar just clear of the forward face of the block. The purpose of this is to prevent the propeller shaft from sliding aft in the event of fracture of an intermediate shaft. Another function of the trailing collar is to allow disconnection of the shaft coupling forward of this to enable free running of the propeller shaft in the event of the ship being towed, due to an engine breakdown, the trailing collar then bears against the forward face of the block.

LOCKING GEAR. A shaft locking gear may be fitted on the shafting so that the shaft can be locked to allow repairs to be carried out on the engine or its transmission system whilst at sea. In twin screw vessels, the opposite engine could continue to run and the locking gear would prevent the free revolving of the propeller of the damaged engine due to the water stream action on the propeller. A simple design of locking gear is the friction type. This consists of a pair of curved flat-section steel straps enveloping one of the couplings, each strap is hinged at its lower end to built-up framework, and a bolt passes through lugs at the top ends of the straps. When not in use, the nut and bolt is tightened up against a distance piece between the lugs which keeps the straps clear of the coupling. To lock the shaft, the distance piece is removed and the nut and bolt tightened until the straps firmly grip the coupling.

PROPELLER SHAFT AND STERN TUBE

The propeller shaft, as its name implies, carries the propeller on its outboard end. It is the last section of shafting and often referred to as the tail-end shaft, or tail shaft. It is coupled to the last intermediate shaft and passes through the stern tube which carries the weight of the propeller shaft and propeller. Referring to Fig. 73, the outboard end of the shaft is tapered and a key with rounded ends sunk into it; the propeller boss fits on this tapered part and the propeller nut screwed on the end to lock the propeller. A locking finger key is fitted on the nut, protruding into a hole in the face of the propeller boss to prevent the nut slackening back. The two usual methods of supporting the propeller shaft in the stern tube are (i) by a lignum vitæ bearing, (ii) by a white metal bearing.

When a lignum vitæ bearing is fitted in the stern tube, the shaft has a brass or bronze liner (sometimes called a sheath) of about 20mm thickness shrunk on it over its parallel length. The bronze liner protects the steel shaft from the corrosive action of the sea water, provides a good surface to run on the lignum vitæ bearing and, when wear takes place, the liner is cheaper to replace than the shaft. A square section recess is cut into the forward face of the propeller boss to take a rubber ring, the after edge of the liner on the shaft presses against this rubber ring to form a seal against sea water coming into contact with the shaft, which would cause corrosion of the steel due to galvanic action.

The stern tube is the cast iron tube through which the propeller shaft passes through the stern of the ship, its function being to support the weight of the propeller and shaft, and prevent water entering the ship. It is fitted between the after peak bulkhead and the stern frame, being placed into position from the inside of the ship. A brass bush is fitted into the stern tube, strips of lignum vitæ wood are fitted inside and held fore and aft by a neck in the bush at the inner end and a brass ring flange at the outer end. The wood strips are bored to suit the diameter of the brass liner on the shaft. This forms the bearing for the shaft and the spaces between the wood strips allow sea water to pass along to lubricate the rubbing surfaces. A stuffing box (with soft greasy packing inserted) and gland is incorporated at the inner end of the stern tube to prevent sea water entering the tunnel.

The wear down of the stern tube bearing is measured regularly, a maximum of 8mm being the usual allowance before the wood strips are renewed. The wear down can be measured by either inserting a small wedge between the shaft liner and the outer end of the bearing

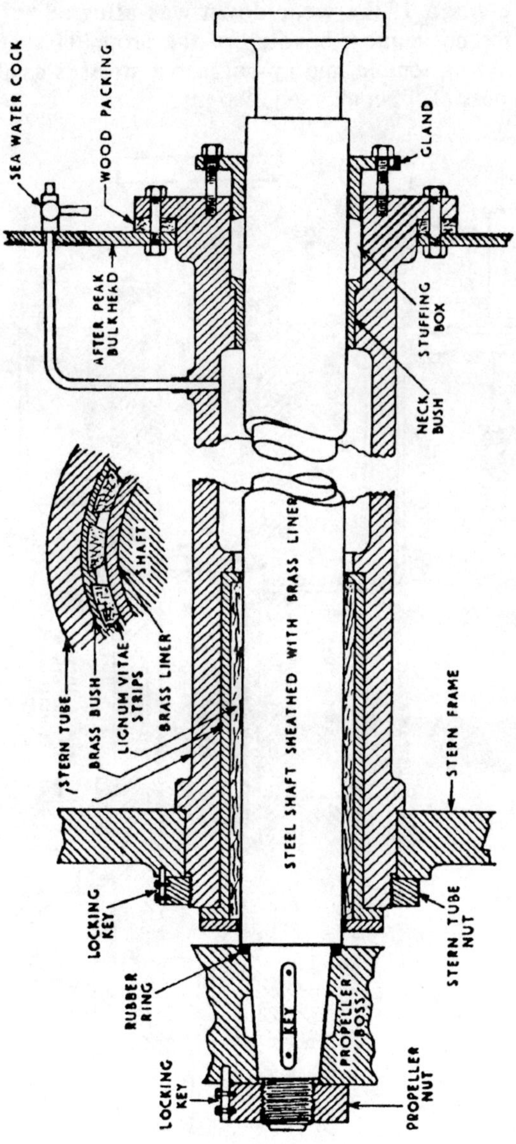

Figure 73
Stern tube and propeller shaft

at the top, or by a poker gauge (having the original marking) inserted down through a hole near the end of the stern tube until it touches the shaft. If the wear down was allowed to become excessive, the consequent bending of the propeller shaft would cause alternating tensile and compressive stresses as the shaft rotates, and possible fracture due to fatigue.

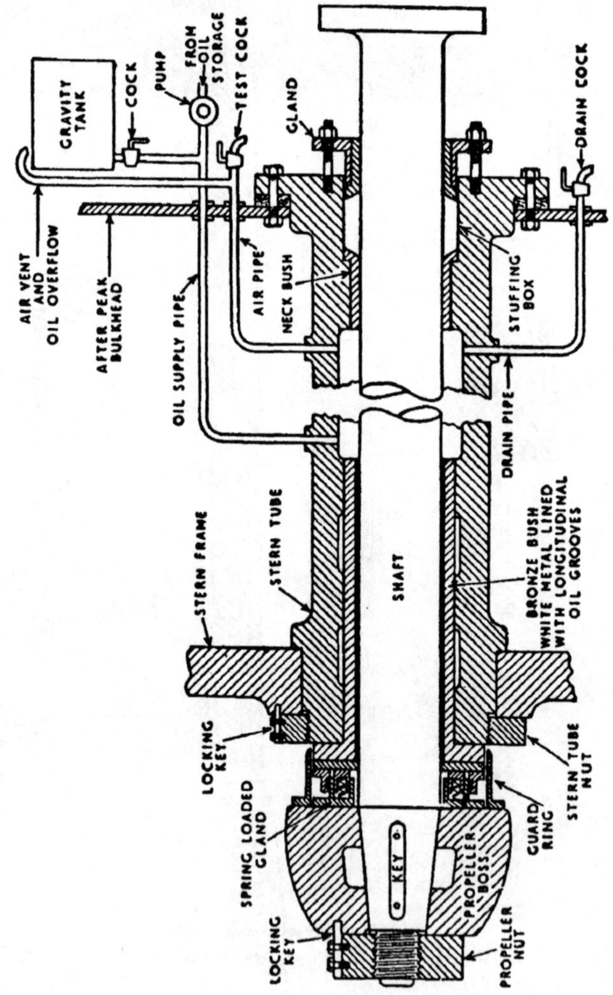

Figure 74
Oil lubricated stern tube

The alternative method is a plain propeller shaft without brass liner, running in a white metal bearing (instead of lignum vitæ) in the stern tube, and lubricated with oil as shown in Fig. 74. To keep the lubricating oil in and the sea water out, a spring-loaded face-to-face gland is fitted at the outer end. This consists of a back plate bolted to the stern tube, and a spring-loaded rotary gland fixed to the propeller boss, so that the gland makes facial contact with the stationary back plate. At the forward end, the gland may be of a similar pattern, or may be of the stuffing box and packing type.

The lubricating oil may be supplied from a gravity tank situated at the required height to maintain the oil in the stern tube at a little above the outside sea water pressure, or it may be pumped in by a pump.

THE PROPELLER

The propeller is a screw of three, four or five blades set at an angle to the plane of rotation so that, when turned, it screws through the water in much the same way as a bolt screws through its nut, and thus converts the engine torque into a direct thrust to push the ship along.

The pitch of a screw is the axial distance it will move forward, working in an unyielding medium, when turned through one revolution.

Referring to Fig. 75, consider a section of a blade. The angle at which it lies to the plane of rotation is termed the *pitch angle* and it can be seen from the development that,

if R = radius from centre of shaft to section,
θ = pitch angle,
then $2\pi R$ = circumference

$$\frac{\text{and pitch}}{2\pi R} = \tan\theta$$

$$\therefore \text{pitch} = 2\pi R \tan\theta$$

Constant pitch propellers are those where every part of the blade has the same pitch, that is, tends to move the same axial distance for one revolution of the propeller. The pitch angle of the blade therefore varies from root to tip such that $R \tan\theta$ is constant, hence the greater the radius of the section from the centre of the shaft, the smaller the angle at that part, and the blades have a twisted appearance when viewed from tip to centre.

Most propellers have, however, a variation of pitch throughout the length of the blade, very carefully designed to take many factors into consideration and resulting in an increased efficiency, the pitch at the root usually being less than the pitch near the tip. The pitch of the propeller is then taken as the average value over the blade length.

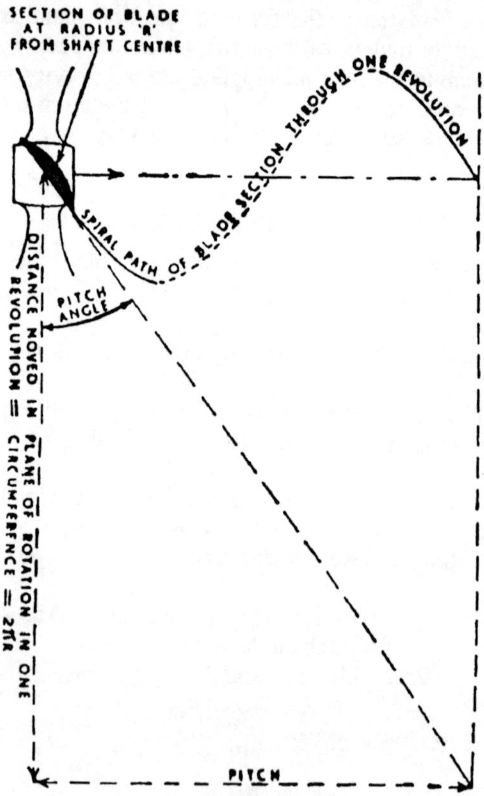

Figure 75
Propeller pitch

MEASUREMENT OF PITCH. This can be done by measuring the pitch angle of the driving face of the blade and noting the radius from the centre to the section at which the pitch angle is measured, then applying the expression, Pitch = $2\pi R \tan \theta$. This should be done at a number of different radii along the blade and the average taken as the propeller pitch.

An instrument called a "pitchometer" is designed on the above principle. It consists of two legs which open out on a hinge like a joiner's folding rule, one leg is held in the plane of rotation (by spirit level or plumb bob) while the other is opened out in line with the blade face. An engraved protractor on the instrument reads the pitch off a table for various combinations of pitch angle and radius.

If such an instrument is not available a weighted cord may be hung over the blade when horizontal as shown in Fig. 76, measuring the distances AC and BC as well as the radius from the centre of the boss. The tangent of the pitch angle is BC divided by AC therefore:

$$\text{Pitch} = 2\pi R \tan\theta$$
$$= \frac{2\pi \times R \times \text{BC}}{\text{AC}}$$

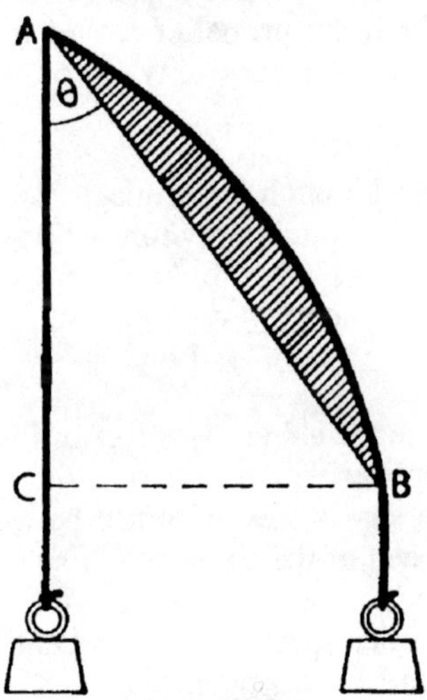

Figure 76
Pitch measurement

Example. The pitch angles of a propeller blade measured at radii of 1, 1.5, 2, 2.5 and 3 metres are respectively 44, 33, 26.5, 22.5 and 20 degrees.

Calculate the pitch of the propeller.
At 1 metre radius
$$\text{pitch} = 2\pi \times 1 \times \tan 44 = 6.067\text{m}$$
At 1.5 metres radius,
$$\text{pitch} = 2\pi \times 1.5 \times \tan 33 = 6.12\text{m}$$
At 2 metres radius,
$$\text{pitch} = 2\pi \times 2 \times \tan 26.5 = 6.265\text{m}$$
At 2.5 metres radius,
$$\text{pitch} = 2\pi \times 2.5 \times \tan 22.5 = 6.506\text{m}$$
At 3 metres radius,
$$\text{pitch} = 2\pi \times 3 \times \tan 20 = 6.861\text{m}$$

$$\begin{aligned}\text{Propeller pitch} &= \text{mean value}\\ &= \underline{6.067 + 6.12 + 6.265 + 6.506 + 6.861}\\ &= 6.364 \text{ metres.}\end{aligned}$$

PROPELLER SLIP. The pitch of a propeller is the distance (in metres) through which the propeller would move forward in one revolution if it worked in a solid unyielding medium. One knot = 1.852km/h, therefore:

$$\begin{aligned}\text{Propeller speed} &= \text{pitch} \times \text{rev/min} && \text{metres/min}\\ &= \text{pitch} \times \text{rev/min} \times 60 && \text{metres/hour}\\ &= \text{pitch} \times \text{rev/min} \times 60 \times 10^{-3} && \text{km/hour}\\ &= \frac{\text{pitch} \times \text{rev/min}}{10^{-3} \times 1.852} \times 60 && \text{knots}\end{aligned}$$

The actual distance the ship moves forward is less than the above because, as the propeller works in water, there is always a certain amount of slip. The slip is the difference between the speed of the propeller and the speed of the ship and is expressed as a percentage of the propeller speed.

Thus, if the propeller speed was equivalent to 16 knots and the speed of the ship was 14 knots then,

$$\text{slip} = 16 - 14 = 2 \text{ knots.}$$
$$2/16 = 1/8 \text{ of propeller speed.}$$
$$1/8 \times 100 = 12\tfrac{1}{2}\% \text{ of propeller speed.}$$

This propeller slip is known as the apparent slip, hence: –

$$\text{Apparent slip} = \frac{\text{Theoretical ships speed} - \text{actual ships speed}}{\text{Theoretical ships speed}} \times 100\%$$

The theoretical ships speed is of course the propeller speed as outlined above.
True slip (or real slip) is the difference between the theoretical ships speed and the speed of advance.
The speed of advance is the speed of the ship relative to the wake.

hence: – True slip = $\dfrac{\text{Theoretical ships speed} - \text{Speed of advance}}{\text{Theoretical ships speed}} \times 100\%$

This value will always be positive as it is independent of the current, which if it was assisting the vessel could lead to a negative Apparent slip.

Example. The pitch of a propeller is 4.9m and the speed of the ship is 13.5 knots when the propeller is turning at 95 rev/min. What is the percentage slip of the propeller?

$$\text{Propeller speed} = \frac{\text{pitch} \times \text{rev/min} \times 60}{10^3 \times 1.852}$$

$$= \frac{4.9 \times 95 \times 60}{10^3 \times 1.852} = 15.09 \text{ knots}$$

$$\text{per cent slip} = \frac{\text{propeller speed} - \text{ship's speed} \times 100}{\text{propeller speed}}$$

$$= \frac{15.09 - 13.5 \times 100}{15.09}$$
$$= 10.54\%$$

VARIABLE PITCH PROPELLER

A variable pitch or *controllable pitch propeller* is of the built-up type with separate boss and blades, the blades being capable of being turned through the required angle to the plane of rotation to change from full pitch angle ahead to full pitch angle astern and be held in any intermediate position while the engine and propeller run in one direction only. Each blade has a circular flange at its root which is secured to, but is a sliding rotational fit in, its corresponding recess in the propeller boss. The flanges are pivoted about their centres and operated by mechanism hydraulically controlled from the ship's bridge.

Thus, manœuvring of the ship ahead or astern, and variation of the ship's speed is done only by controlling the pitch of the propeller blades. Some of the advantages achieved by this propeller system are:

1. Improved manœuvrability.
2. Uni-directional engine, no reversing mechanism required.

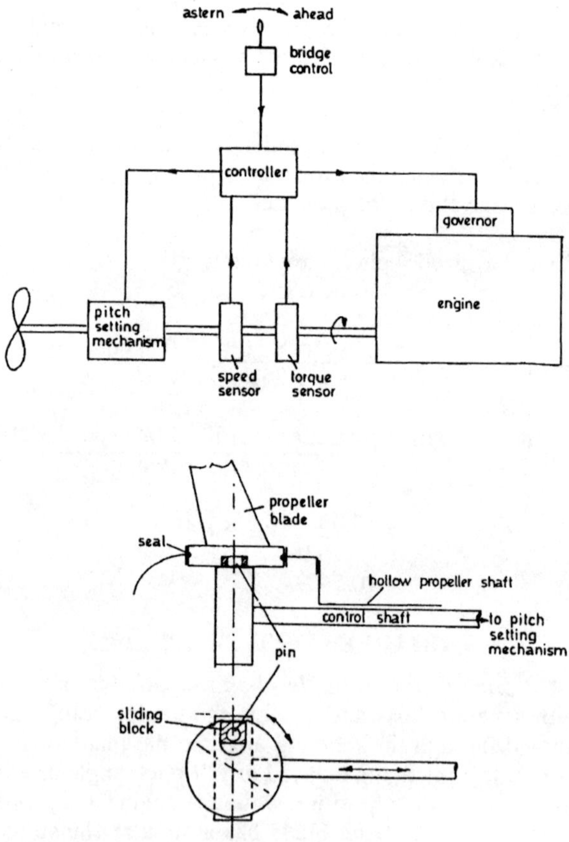

Figure 77
Controlled pitch propeller

3. Reduced number of engine starts.
4. Improved engine efficiency.
5. Engine room personnel freed from stand-by duties.

Disadvantages include:

1. Complex propeller system hence reduced reliability compared to fixed pitch.
2. With oil in propeller boss possibility of pollution of the sea with oil in event of seal failure.
3. Increased dry docking time for propeller survey.

Fig. 77 outlines the pneumatic or electronic control where a single lever on the bridge (a duplicated system exists in the engine room) is used to send the desired vessel direction and speed to the controller. By controlling pitch, engine speed and torque the best possible economic use can be obtained.

FRICTION OF SHIPS' HULLS THROUGH WATER

The amount of friction between the hull of the ship and the water through which it is moving is proportional to approximately the square of the speed, to the wetted surface area, and to the density of the water; it also depends upon the roughness of the surface of the shell; it is independent of the water pressure, which means that there is the same amount of friction per square foot at the bottom of the hull as there is near the water surface.

From the rule that areas of similar figures vary as the square of their corresponding dimensions, we have,

Wetted surface area varies as $(\text{length})^2$

From the rule that volumes of similar objects vary as the cube of their corresponding dimensions, we have,

Displacement varies as $(\text{length})^3$

Therefore, length varies as $(\text{displacement})^{1/3}$
hence, Wetted surface area varies as $(\text{displacement})^{2/3}$
Since Friction varies as wetted surface area and $(\text{speed})^2$
then Friction varies as $(\text{displacement})^{2/3}$ and $(\text{speed})^2$

$$\text{Power} = \text{force} \times \text{speed}$$

$$\therefore \text{Power varies as } (\text{displacement})^{2/3} \times (\text{speed})^2 \times \text{speed}$$
$$= (\text{displacement})^{2/3} \times (\text{speed})^3$$

The power of the engines is proportional to the power to propel the ship through the water, therefore, engine power varies as (displacement)$^{2/3}$ x (speed)^3

Hence $\dfrac{\text{displacement}^{2/3} \times \text{speed}^3}{\text{engine power}} = \text{a constant}$

This constant is called the Admiralty Coefficient and may be used for estimating the power of the ship's engines under different conditions of displacement and speed, or for the comparison of the engine power of similar ships.

If Δ = displacement in tonne
V = speed in knots
P = power

$$\frac{\Delta_1^{2/3} \times V_1^3}{P_1} = \frac{\Delta_2^{2/3} \times V_2^3}{P_2}$$

Example. The power developed by a ship's engine was 3,200kW when her displacement was 10,000 tonne and speed 14 knots. Estimate the power required to run at a speed of 16 knots when her displacement is 12,000 tonne.

$$\frac{\Delta_1^{2/3} \times V_1^3}{P_1} = \frac{\Delta_2^{2/3} \times V_2^3}{P_2}$$

$$\frac{10{,}000^{2/3} \times 14^3}{3{,}200} = \frac{12{,}000^{2/3} \times 16^3}{P_2}$$

$$P_2 = \frac{3{,}200 \times 12{,}000^{2/3} \times 16^3}{10{,}000^{2/3} \times 14^3}$$

$$P_2 = 3{,}200 \times \left\{\frac{12{,}000}{10{,}000}\right\}^{2/3} \times \frac{16^3}{14}$$

$$P_2 = 5{,}394\text{kW}$$

Many cases will arise when the power is required to be estimated for change of speed only, that is, assuming that the displacement remains the same or has such little change of displacement that it can be neglected.

In such cases (displacement)$^{2/3}$ cancels both sides to leave

$$\frac{\text{speed}_1^3}{\text{power}_1} = \frac{\text{speed}_2^3}{\text{power}_2}$$

FUEL CONSUMPTION

The mass of fuel burned in a given time (per hour or per day) is proportional to the power developed by the engines, therefore:

$$\frac{(\text{displacement})^{2/3} \times (\text{speed})^3}{\text{daily consumption of fuel}} = \text{a constant}$$

This constant is termed the Fuel Coefficient and by the use of this expression the quantity of fuel required under different conditions of displacement and speed can be estimated, thus:

$$\frac{\Delta_1^{2/3} \times V_1}{\text{daily consumption}_1} = \frac{\Delta_2^{2/3} \times V_2^3}{\text{daily consumption}_2} \quad ....(1)$$

and this may be written in the form:

$$\frac{\text{new daily cons.}}{\text{old daily cons.}} = \left\{\frac{\text{new displ.}}{\text{old displ.}}\right\}^{2/3} \times \frac{\text{new speed}^3}{\text{old speed}}$$

$$\text{Daily consumption} = \frac{\text{consumption over the whole voyage}}{\text{number of days on voyage}}$$

$$\text{Number of days on voyage} = \frac{\text{distance}}{\text{speed in knots} \times 24}$$

$$\therefore \text{ daily consumption} = \frac{\text{voyage consumption} \times \text{speed} \times 24}{\text{distance}}$$

Substituting for daily consumption, new and old, into equation (i) we get:

$$\frac{\Delta_1^{2/3} \times V_1^3 \times \text{distance}_1}{\text{voyage consumption}_1 \times V_1 \times 24} = \frac{\Delta_2^{2/3} \times V_2^3 \times \text{distance}_2}{\text{voyage consumption}_2 \times V_2 \times 24}$$

V_1 cancels into V_1^3 to make it V_1^2
V_2 cancels into V_2^3 to make it V_2^2
24 cancels from each side.

Therefore the equation may now be written:

$$\frac{\text{New voy. cons.}}{\text{Old voy. cons.}} = \left\{\frac{\text{New displ.}^{2/3}}{\text{Old displ.}}\right\} \times \left\{\frac{\text{New spd.}}{\text{Old spd.}}\right\}^2 \times \frac{\text{New dist.}}{\text{Old dist}}$$

The above expression can be taken as the general formula for solving most fuel consumption problems.

The voyage can be taken as the day's run where the consumption of fuel per day is given, thus the daily fuel consumption can be put down as the voyage consumption if the corresponding distance be taken as speed (in knots) × 24.

If there is no change in displacement, this term will cancel out.

If the distance is the same, this term will also cancel.

The worked examples to follow will clarify these few points.

Example. A vessel uses 300 tonne of fuel on a voyage of 3,000 nautical miles travelling at a speed of 12 knots when her displacement is 10,000 tonne. Estimate the fuel required for a voyage of 1,500 nautical miles at a speed of 15 knots when her displacement is 14,000 tonne.

$$\frac{\text{New voy. cons.}}{\text{Old voy. cons.}} = \left\{\frac{\text{New displ.}}{\text{Old displ.}}\right\}^{2/3} \times \left\{\frac{\text{New spd.}}{\text{Old spd.}}\right\}^{2} \times \frac{\text{New dist.}}{\text{Old dist.}}$$

$$\frac{\text{New voy. cons.}}{300} = \left\{\frac{14{,}000}{10{,}000}\right\}^{2/3} \times \left\{\frac{15}{12}\right\}^{2} \times \frac{1{,}500}{3{,}000}$$

$$\text{New voy. cons.} = 300 \times 1.4^{2/3} \times 1.25^{2} \times 0.5$$

$$= 293.3$$

Example. A vessel of 10,000 tonne displacement burns 25 tonne of fuel per day when her speed is 12 knots. Calculate the probable consumption of fuel over a voyage of 3,000 nautical miles at a speed of 11 knots when the displacement is 11,000 tonne.

The original fuel consumption is given as 25 tonne per day therefore assume the original voyage to be one day's run which is 12 × 24 nautical miles, and the consumption of fuel for the voyage as 25 tonne.

$$\frac{\text{New voy. cons.}}{\text{Old voy. cons.}} = \left\{\frac{\text{New displ.}}{\text{Old displ.}}\right\}^{2/3} \times \left\{\frac{\text{New spd.}}{\text{Old spd.}}\right\}^{2} \times \frac{\text{New dist.}}{\text{Old dist.}}$$

$$\frac{\text{New voy. cons.}}{25} = \left\{\frac{11{,}000}{10{,}000}\right\}^{2/3} \times \left\{\frac{11}{12}\right\}^{2} \times \frac{3{,}000}{12 \times 24}$$

$$\text{New voy. cons.} = \frac{25 \times 1.1^{2/3} \times 121 \times 3{,}000}{144 \times 12 \times 24}$$

$$= 233.1 \text{ tonne.}$$

Example. A ship travels 900 nautical miles at a speed of 12.5 knots and burns 150 tonne of fuel over the voyage. Estimate the distance the ship could travel at a speed of 13.5 knots on 250 tonne of fuel.

Assuming the displacement to be the same in each case:

$$\frac{\text{New voyage consumption}}{\text{Old voyage consumption}} = \left\{\frac{\text{New speed}}{\text{Old speed}}\right\}^2 \times \frac{\text{New distance}}{\text{Old distance}}$$

$$\frac{250}{150} = \frac{13.5^2}{12.5^2} \times \frac{\text{New distance}}{900}$$

$$\text{New distance} = \frac{250 \times 12.5^2 \times 900}{150 \times 13.5^2}$$

$$= 1{,}286 \text{ nautical miles.}$$

CHAPTER TEN

CONTROL FUNDAMENTALS

Control systems are either automatic or remote or a combination of these. Early automatic controls were self operating, which means that the power to operate the control unit was supplied direct by a change in condition of the medium under control. Examples of these are the safety valve or a float controlled valve for liquid level as shown in Fig. 78. In Fig. 78a a change in demand results in a change in position of the float which in turn alters the valve position until supply once again matches demand.

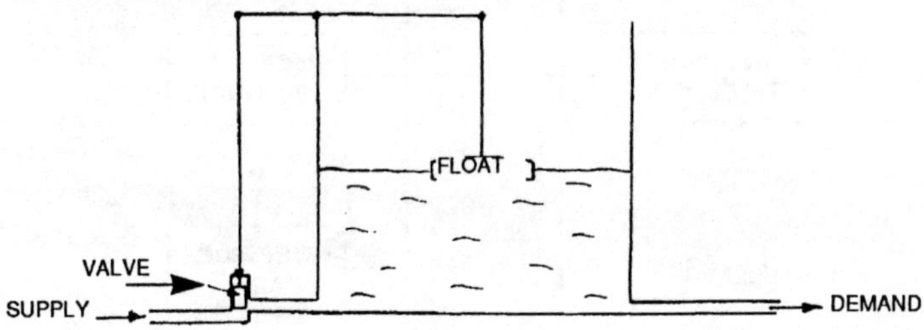

Figure 78a

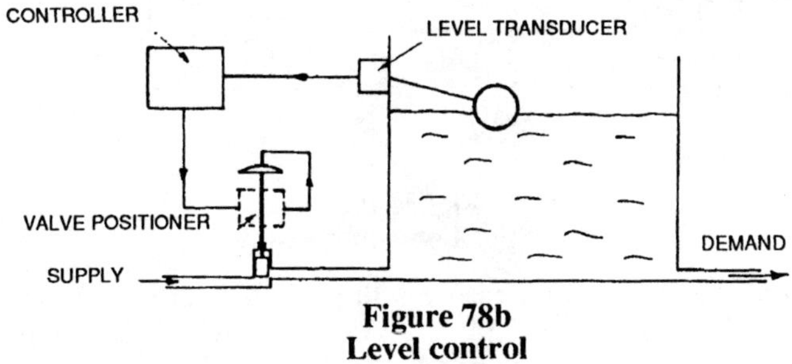

Figure 78b
Level control

Fig. 78b shows the automatic liquid level control in its modern form. The level of the liquid would be sensed by a sensor – this may be a float, pressure device or capacitor – which then has its message transduced into an electric or pneumatic signal that is relayed to the controller.

If the signal received by the controller is in agreement with a desired value that the controller is set at, no alteration in the controlled valve position will take place. However, if the signal received by the controller is not in agreement with the desired value, the controller sends a signal to the valve positioner to re-position the valve in order that supply can be made to match demand.

With this arrangement the valve and position control unit can be remote from the tank, level can be controlled accurately and reliably.

Remote control means that the system is being controlled by the operator who is situated remote from the system. Examples are: steering gear, water tight door system, tanker cargo control, bridge control of engines, etc.

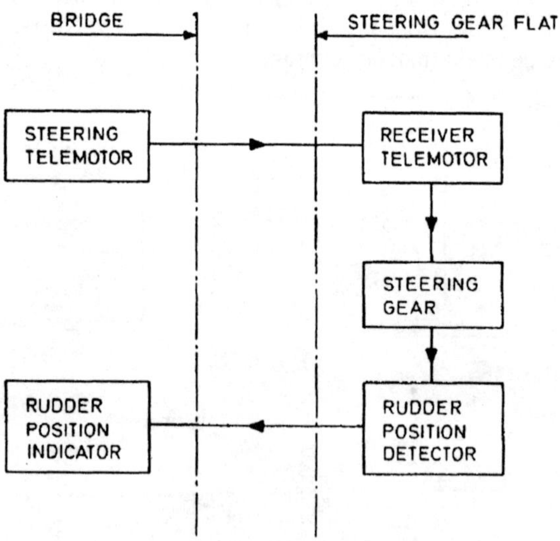

Figure 79
Remote position control system

In order to acquaint the reader with modern control terminology the excellent example of the steering gear, which is a remote position control system, will be examined.

Fig. 79 shows diagrammatically the arrangement which would be operated as follows. The operator positions the steering telemotor at the *desired value* of rudder position, which *transduces* his movement into a signal which is transmitted to the receiver telemotor, which then commands the steering gear to position the rudder at the *desired value.* A rudder position indicator *feeds back* a signal to a display unit on the bridge, this *measured value* would be compared by the operator with the *desired value* and any *deviation* would cause him to alter the position of the steering telemotor.

Replacing the operator by means of an auto-helmsman converts this *remote position closed loop control system* into an *automatic closed loop control system.*

All the foregoing terms in italics together with others in common use in control systems will now be defined.

Automatic open loop control system is one in which the control action is independent of the output, that is, the output in no way controls the system. For example a coffee percolator, what is its output? the colour of the coffee? strength of the coffee? temperature? flavour? none of them are being used to exercise control over the percolator action. The only control is time based, we put in the ingredients, set the timer, switch on and then after a predetermined time has elapsed the percolator automatically switches itself off.

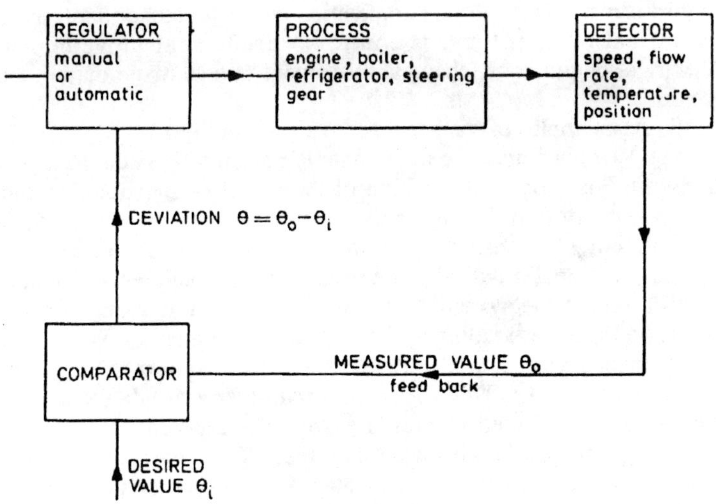

Figure 80
Closed loop control system

There are many other examples, washing machine, toaster; marine examples are the oil purification system and soot-blowing system.

Closed loop control system is one in which the control action is dependent on the output. The system may be manually controlled or automatic, examples being steering gear, boiler control, engine control, cargo control, etc. Fig. 80 shows the basic elements in a closed loop control system.

The measured value of the output is being fed back to the controller which compares this value with the desired value for the controlled condition and produces an output to alter the controlled condition if there is any deviation between the values. *Measured Value*; actual value of the controlled condition (symbol θ_0).

Desired Value (or set value); the value of the controlled condition that the operator desires to obtain. Examples, 2 rev/s, 25 degrees of helm, 55 bar, –5°C, etc. (symbol θ_i).

Deviation (or error); is the difference between measured and desired values (symbol θ). Hence $\theta = \theta_0 - \theta_i$ This signal, probably converted into some suitable form such as voltage to hydraulic output or voltage to pneumatic output, etc., would be used to instigate corrective action – object to reduce error to zero.

Offset; is sustained deviation.

Transducer; changes the signal received from the detector/sensor into some readily amplified output, usually electrical. The term transducer has now general application and is no longer confined to electrical terminology, it could be mechanical movement to electrical output or mechanical movement to pneumatic output, and so on.

Some examples of transducers are shown in Fig.81.

Fig. 81(a) is a pressure to pneumatic transducer. As the measured pressure Pm varies, the position of the end A of the Bourdon tube also varies. In turn this alters the position of the flapper and hence the amount of air escaping from the nozzle will change. Output signal pressure Po will alter and the feedback bellows will restore equilibrium to the system so that we now have a new output signal pressure Po corresponding to the new measured pressure Pm.

A novel pressure to electrical transducer is shown in simplified form in Fig. 81(b). This uses a tourmaline crystal whose opposite faces when subjected to pressure produces a potential difference (this is known as the piezoelectric effect). This potential difference can be sensed, amplified and displayed. This type of transducer is suitable for varying pressures *e.g.*Diesel engine cylinder pressures.

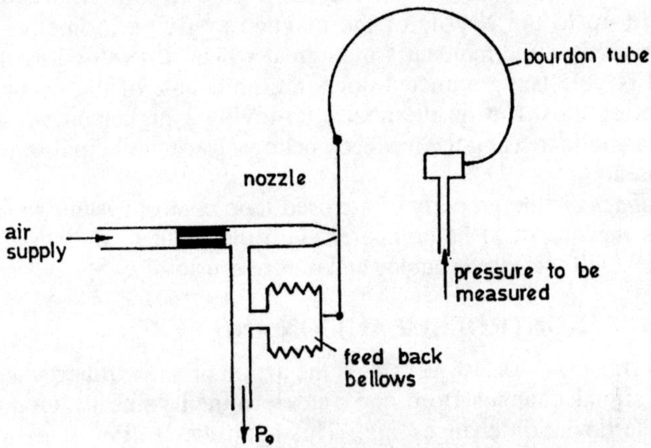

Figure 81a

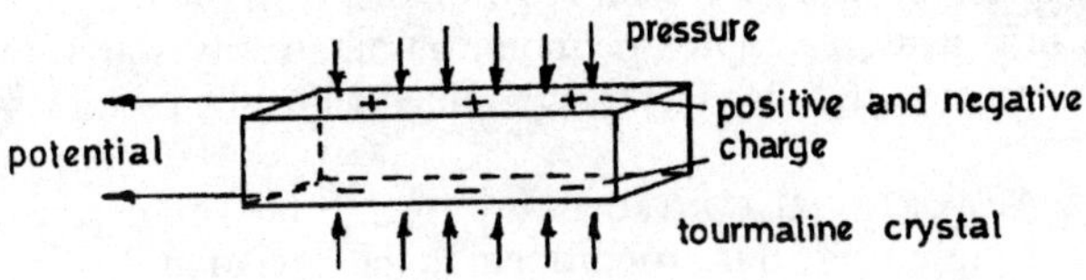

Figure 81b

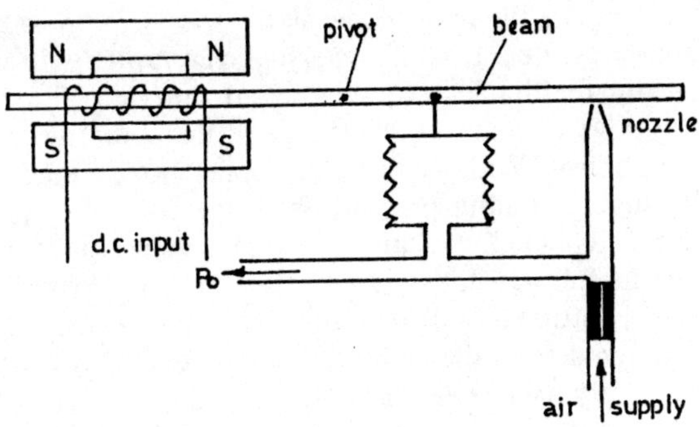

Figure 81c

An electric to pneumatic transducer is illustrated in Fig. 81(c). For an increase in current the created S pole to the left will be attracted up to the N pole of the magnet so giving a clockwise rotation because the moment arm is greater than that caused by the created N pole being attracted down on the S pole of the magnet. This action closes in on the nozzle so giving a higher output air pressure and increasing the feedback bellows force until equilibrium is restored.

Feedback; is the property of a closed loop control system which permits the output to be compared with the input to the system. Feedback will increase accuracy and reduce sensitivity.

CONTROLLER ACTION AND TYPES

1. TWO STEP CONTROLLER ACTION; is the action of a controller whose output signal changes from one predetermined value to another when the deviation changes sign. This controller action is mainly on/off control, for example, a refrigeration unit controlling room temperature, when the temperature rises to a predetermined value the refrigerator compressor motor automatically starts, then when the temperature falls to a predetermined value the motor is stopped.

2. PROPORTIONAL CONTROLLER ACTION; is the action of a controller whose output signal is proportional to the deviation.

Using Fig. 82, if h_i is the desired head of fluid in the tank and the demand is reduced so that the head increases to h_0 then the deviation in head of fluid is $h_0 - h_i$. The level sensor-transducer produces a signal proportional to the deviation in head, that is, an air pressure increase which will deflect the bourdon tube an amount θ reducing flapper nozzle separation in the controller.

The output pressure p from the pneumatic controller will increase and act on the feedback bellows B_1 and the valve positioner. Feedback bellows B_1 increases the range of the controller *ensuring* proportional control. Without it, since flapper nozzle separation is small, full range of output pressure from the controller would be traversed for a very small deviation θ and this would cause the valve to move over its full travel, that is, on/off or two step control.

The valve positioner will position the valve according to the command received from the controller, reducing fluid flow into the tank so that supply once again matches demand.

However we now have a steady head condition in the tank, h_0 which is $h_0 - h_i$ away from the desired head, in other words we have *offset* – sustained deviation.

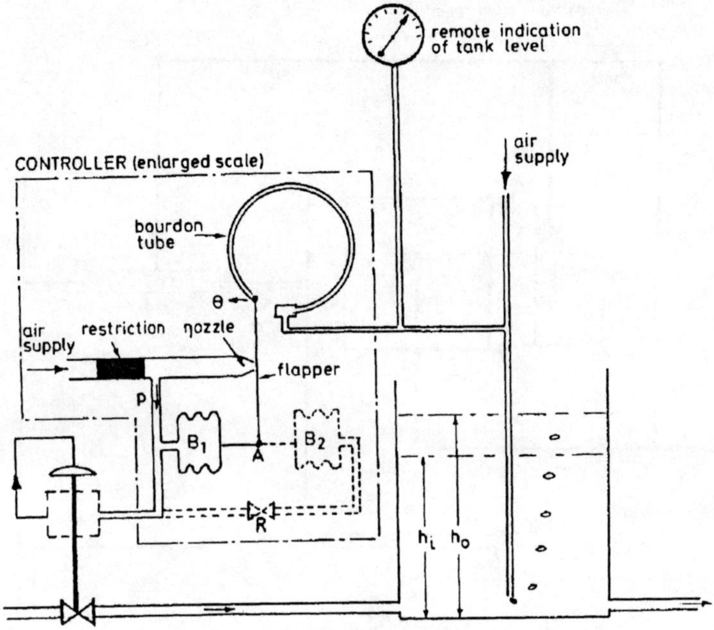

Figure 82
Pneumatic, proportional and proportional with integral control

3. INTEGRAL CONTROLLER ACTION; is the action of a controller whose output signal changes at a rate which is proportional to the deviation. The object of integral control action is to reduce offset to zero, hence integral action is usually called reset action.

In Fig. 82 the controller shown in full lines gives proportional control with offset. If the bellows B_2 and the adjustable restrictor R are connected into the system as shown by the dotted lines the controller has now reset action as well as proportional action.

If p increases as before, point A will move to increase flapper nozzle separation, however the air pressure in B_2 will rise (at a rate depending upon the setting of the restriction R) retarding the movement of the point A.

Fig. 83 shows a proportional and integral controller for the hydraulic – mechanical positioning of a controlled valve.

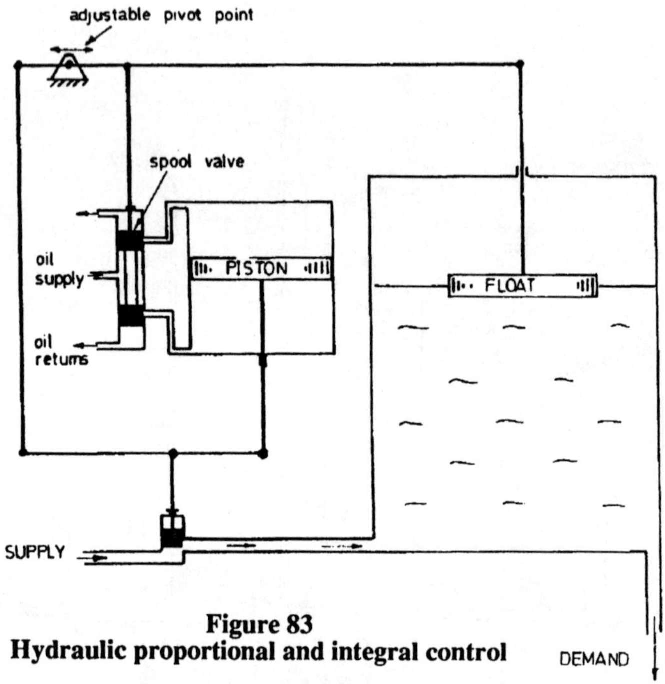

Figure 83
Hydraulic proportional and integral control

If the float is rising above the set value due to a reduction in demand the spool valve will move up and oil under pressure will act on the piston top to close the controlled valve, the mechanical linkage will also act to close the valve.

For a falling float above the set value the actions of the piston and mechanical linkage are in opposition. For a falling float below set value both act to open the valve.

Fig. 84 shows a controller of the electronic type for proportional and integral control.

Considering proportional action only; the deviation θ causes movement of the adjustable rheostat at A so that the input bridge is unbalanced and a voltage difference exists between A and B and current flows to the amplifier. Current then flows to the electric motor causing the motor to move the controlling element (*e.g.*, positioning a valve) and as it does so it moves the adjustable rheostat at X so that X and Y now have a voltage difference, that is, the output bridge is unbalanced. The unbalance on the output bridge produces a current in opposition to the current produced by the input bridge and if these opposing currents are equal the voltage across the amplifier is zero and the motor stops.

For each position of A there is a corresponding proportional position for X, this proportionality can be adjusted by the variable resistance R_1.

To obtain proportional and integral action the capacitor C and resistance R are necessary. Variable resistance R can be adjusted to give the required reset action time.

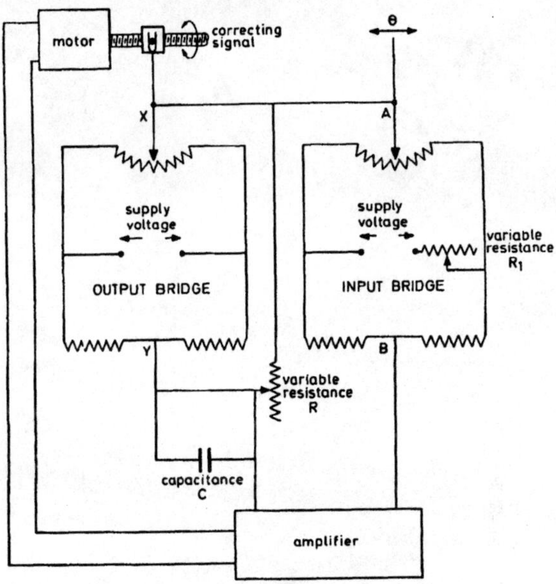

Figure 84
Electronic proportional and integral controller

4. DERIVATIVE CONTROLLER ACTION: is the action of a controller whose output signal is proportional to the rate at which the deviation is changing.

The principal object of derivative control is to give quicker response to system changes.

Sophisticated controllers for rapid response, minimum offsets and reset action would be three term controllers, *i.e.*, proportional plus integral plus derivative.

CHAPTER ELEVEN

CONTROLLED SYSTEMS AND INSTRUMENTATION

Marine control systems are either pneumatic, hydraulic or electronic, or a combination of these. The electronics could act as the nervous system with the pneumatics or hydraulics supplying the muscle. The advantages of the systems will now be examined and it will be left as an exercise for the reader to list the disadvantages.

PNEUMATIC SYSTEM ADVANTAGES:

1. Less expensive than electronic or hydraulic systems.
2. Leakages are not dangerous.
3. No heat generation, hence no ventilation required.
4. Reliable.
5. Not very susceptible to variations in ships power supply.
6. Simple and safe.

ELECTRONIC SYSTEM ADVANTAGES:

1. Fewer moving parts hence less lubrication and wear.
2. Low power combustion.
3. System is either on or off, with pneumatic or hydraulic systems if they develop a leak, or dirt enters the system, they become sluggish.
4. Small and adaptable.
5. Very quick response.

HYDRAULIC SYSTEM ADVANTAGES:

1. Nearly instant response as fluid is virtually incompressible.
2. Can readily provide any type of motion such as reciprocating or rotary.
3. Accurate position control.

4. High amplification of power.

ENGINE ROOM CONTROL SYSTEMS

Apart from the main engine control systems there are many subsidiary automatic control loops. For a diesel engined installation these would include:

Lubricating oil pressure and temperature control.
Fuel oil viscosity control.
Fuel oil purification.
Cooling water pressure and temperature control for jackets, pistons and valves.
Auxiliary boiler control.
Bilge level.

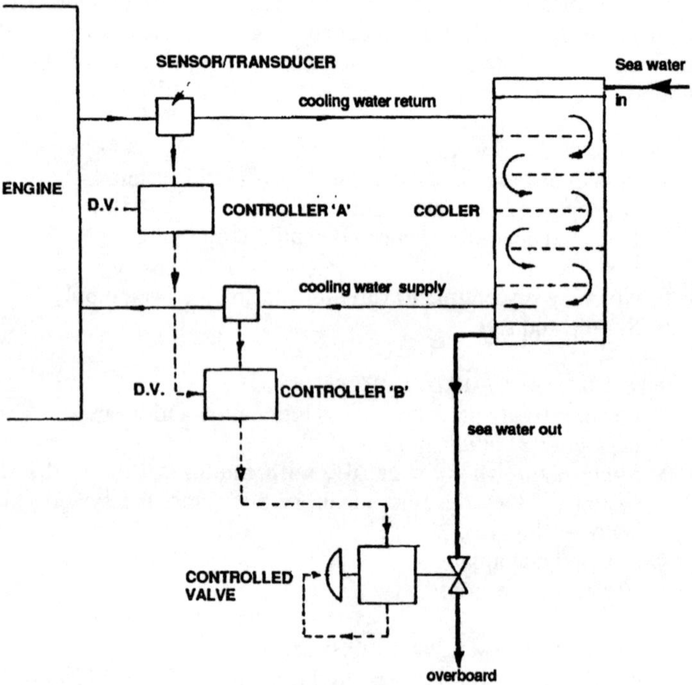

Figure 85
Cascade control

An example of one of these control loops is shown in Fig. 85. This is an automatic closed loop cascade control system for the cooling water used in a diesel engine. Two variables are involved, engine load and sea water temperature. First consider fixed engine load, then controller B senses changes in temperature of cooling water supply and adjusts the sea water flow through the cooler to keep cooling water inlet temperature to the engine at a desired value.

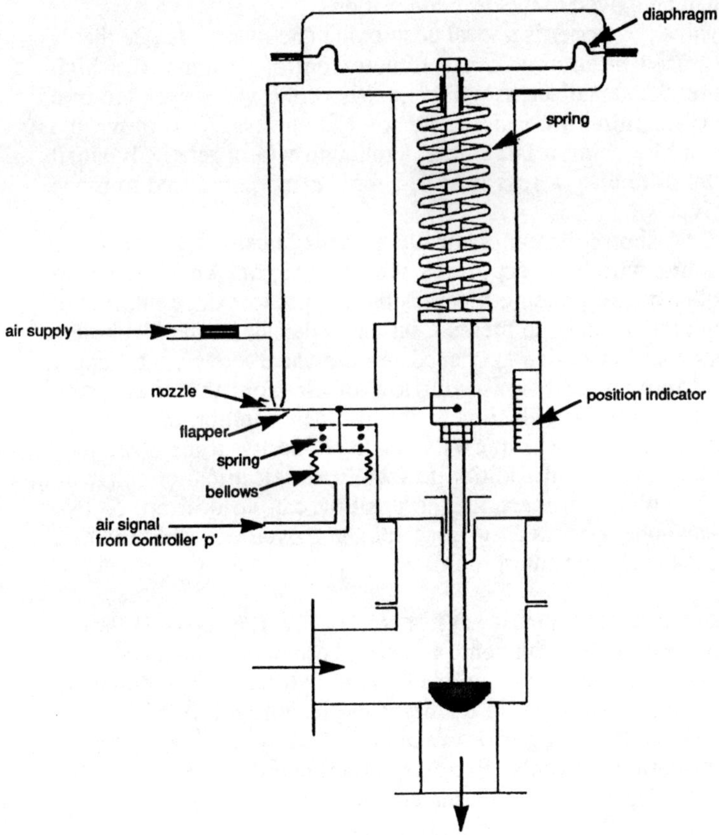

Figure 86
Valve positioner

Now, if the engine load changes, inlet water temperature should change or the engine may become overheated or over-cooled. If we assume the engine load decreases then the cooling water outlet temperature would decrease, controller A senses this change and alters the desired value of Controller B to a higher temperature setting. This type of control is called cascade control, controller A is the master controller and B the slave.

In the foregoing cascade control system the slave controller B sends a command signal to a valve positioner, the controlled valve will then be moved to the desired position.

A valve positioner is a local positioning and muscle device that is widely used in automatic and remote control systems. If a high pressure difference across a valve exists or the valve operated in a viscous medium, a considerable force is necessary to move the valve and by using a large area diaphragm with a relatively small pressure difference across it a large force can be produced to move the valve.

Fig. 86 shows diagrammatically a pneumatically operated valve positioner which incorporates the diaphragm. An increase in controller output pressure p causes the flapper to reduce outflow of air from the nozzle, the pressure on the underside of the diaphragm increases and the valve is opened. As the valve moves, the flapper will be moved to increase outflow of air from the nozzle and eventually the valve will come to rest in a new equilibrium position.

Position indication of the valve is given locally. If the position is required remotely, in addition to locally, an electrical circuit with possibly voltage proportional to position can be utilised, or two micro-switches operated by valve spindle movement could indicate open or shut in the control room.

ELECTRICAL SERVO-MOTORS (ACTUATORS) d.c The servo-motor is a conventional motor with control of field current or armature voltage by a controlling device. Low inertia and high torque is required so that armatures are small in diameter and lengthened. A wide range of speeds is possible, performance is limited by heating caused by high armature currents. Reversal is achieved by reversing the current through the field or armature via the controlling device.

a.c. Three phase induction is cheap and reliable and thyristor circuitry is best for control. Starting torque is lower than d.c. but it can be improved by using high resistance rotors, which unfortunately generates heat. Two phase is used in low power systems for position control *e.g.* pen recorders, potentiometers.

FAIL SAFE

This is a term frequently used in control, it means that if control fails the system under control will adopt a non-dangerous state. It may be shut down or it may continue with fault indication given to the operators. For example if the control air supply to a boiler fuel regulating valve stopped due to air pipe breakage or blockage, the fuel regulating valve would be closed by the spring force preventing un-regulated oil flow to the boiler burners and a possible dangerous condition arising. Hence the "fail safe" position for the boiler fuel regulating valve is "shut".

Some fail safe positions are however open to argument. If a controllable pitch propeller fail safe position was with the propeller blades in the ahead position and the vessel was waiting in a lock, with engines running – propeller blades in neutral, for the lock gates to open and the hydraulic control failed the propeller blades now suddenly adopt their fail safe position! Enough said.

BOILER WATER LEVEL CONTROL

Sometimes simply referred to as feed water control. This may take one of three arrangements. Single element, two element or three element control. The elements are the measuring elements, water level, steam flow and feed flow. Which arrangement chosen depends upon the boiler and plant rating. If the plant is highly rated, *i.e.* with a high output, high pressure water tube boiler the three element control system would be chosen. As an example the two element control system will be described, this would be suitable for moderately rated plant. Two element control. Elements are steam flow and water level measuring each of which sends out a pneumatic signal proportional to changes in its measured variable. Fig. 87. The steam flow signal passes to the relay which transmits a signal to the valve positioner, via the hand auto unit, which alters the valve position an amount proportional to the steam flow. By correct adjustment to the proportional band in the relay and correct setting up of characteristic in the valve positioner it is possible to match changes of load with feed. The water level signal is fed into a P+I controller whose set value is adjusted for desired level of boiler water. Any deviation between measured and desired values results in a change in output signal from the controller to the relay. This signal is added to the steam flow signal and is used to correct deviation in level which could occur due to unbalance of boiler feed in and steam flow out.

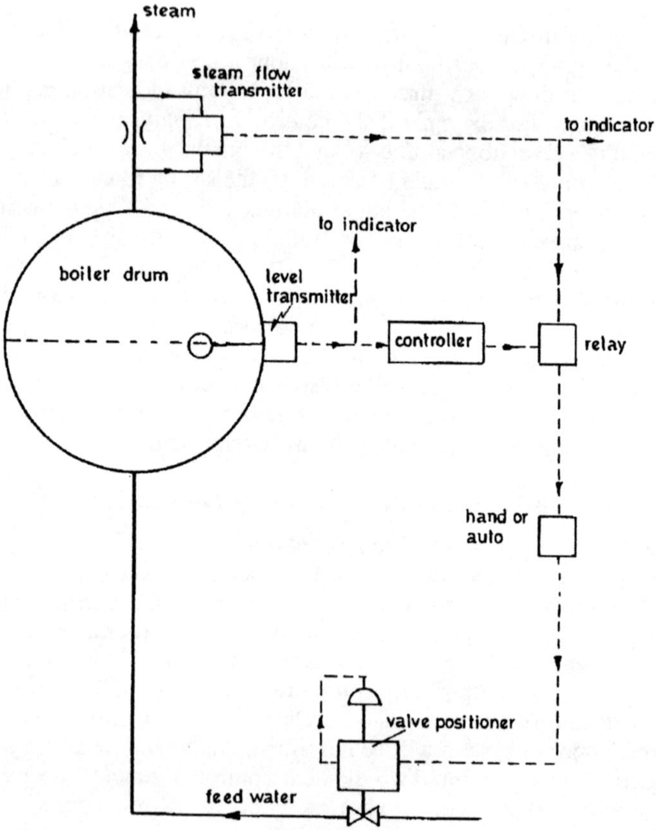

Figure 87
Two element feed water control

U.M.S. Regulations. These may be summarised as follows: –

1. Bridge control of propulsion machinery. The bridge watchkeeper must be able to take emergency engine control action. Control and instrumentation must be as simple as possible.
2. Centralised control and instruments are required in machinery spaces. Engineers may be called to the machinery spaces in emergency and controls must be easily reached and fully comprehensive.
3. Automatic fire detection system. Alarm and detection system

must operate very rapidly. Numerous well sited and quick response detectors (sensors) must be fitted.

4. Fire extinguishing system. In addition to conventional hand extinguishers a control fire station remote from the machinery space is essential. The station must give control of emergency pumps, generators, valves, ventilators, extinguishing media, etc.
5. Alarm system. A comprehensive machinery alarm system must be provided for control and accommodation areas.
6. Automatic bilge high level fluid alarms and pumping units. Sensing devices in bilges with alarms and hand or automatic pump cut in devices must be provided.
7. Automatic start emergency generator. Such a generator is best connected to separate emergency bus bars. The primary function is to give protection from electrical blackout conditions.
8. Local hand control of essential machinery.
9. Adequate settling tank storage capacity.
10. Regular testing and maintenance of instrumentation.

BRIDGE CONTROL OF ENGINES

The control of main propulsion systems and bow thrust units from the bridge has advantages, some are: (i) Response to bridge order is quicker and more consistent, this facilitates manœuvring operations. (ii) More accurate speed control is possible. (iii) Engineers are freed from positions at the controls.

Unattended machinery spaces (designated u.m.s. in the regulations) are, on modern automated vessels, mainly unattended overnight and it is essential therefore that during this time the bridge officer on watch should have control of the main engines.

In the event of minor emergencies when the machinery space is unattended, the engineer on cabin watch will be informed of malfunctions in the machinery by his cabin monitor, which gives audible and visual alarm of such things as incorrect lubricating oil pressure or temperature, incorrect cooling water pressure or temperature, high water level in bilges, etc.

Fig. 88 shows a typical bridge control unit for a diesel engined vessel. In order to operate, the control selection switch would be turned to bridge control and if the over-riding controller in the engine room has been switched to bridge control the bridge officer can operate the engine direct. Let us assume the speed control lever is moved to about 100 rev/min ahead, the following sequence of events would automatically occur:

1. Providing the turning gear is out, fuel off, lubricating oil on, cooling water on and starting air lever in ahead position, starting air would be supplied to the engine.
2. When the engine reaches firing speed the starting air will be shut off.
3. If the engine is turning in the correct direction at the correct speed fuel will be supplied.
4. The fuel supply will be gradually increased to bring engine speed up to the desired value.

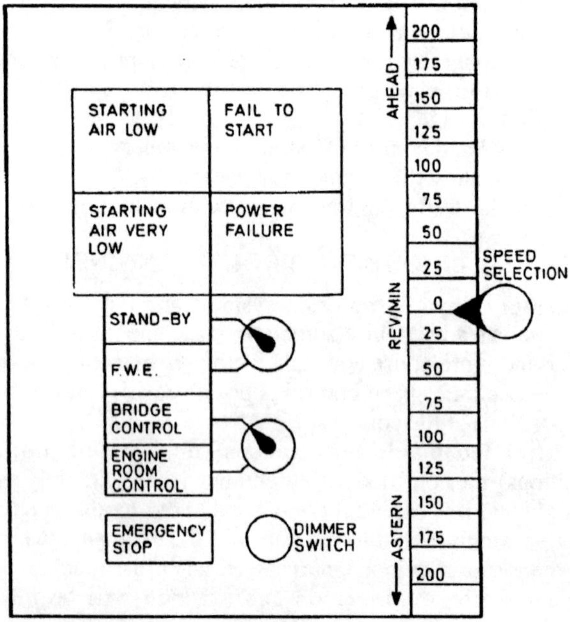

Figure 88
bridge speed unit control

If the engine should fail to start, alarm indication would be given and the operational sequence could be repeated. Stalling of the engine is prevented by the engine governor if the speed control setting is low for the prevailing conditions. However it is strongly recommended that slow speed running for prolonged periods be avoided.

TURBINE MACHINERY. Fig. 89 shows a simplified control system for turbine machinery. The fundamental principles involved are similar to the diesel engined installation and essential interlocks instrumentation and alarm provision has to be made.

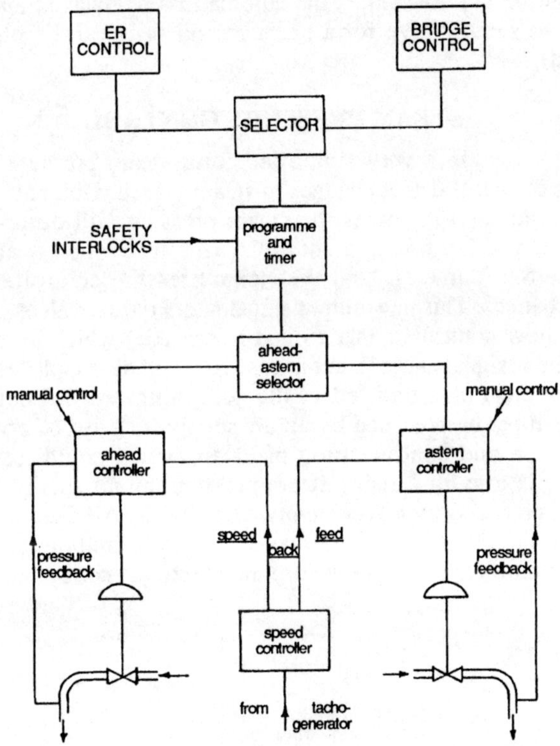

Figure 89
turbine control system

Again, assuming the speed control lever on the bridge is moved to say 100 rev/min ahead, then providing turning gear is out, steam raised on the boiler, condenser circulating water on, lubricating oil on, turbines warmed through, gland steam on, etc., the ahead steam valve will be automatically opened. The rate of opening of the valve is controlled, as sudden opening or closing would give insufficient time for the boiler to adjust itself to the new conditions, this is the duty of the programme timer.

If the system is on manœuvring, a mode switch would be in the on position – this opens the astern guard valve, shuts off the bled steam to heaters, operates main circulating pump at high speed and opens I.p. turbine drains if the speed falls below a certain value. When the engines are stopped during manœuvring, the turbines would be on auto-blast, *i.e.*, the automatic time delayed opening of the ahead steam valve for a short period (this can be blocked if required).

STEAM PRESSURE CONTROL

Fig. 90 shows in a very simplified form steam pressure control. Sensing elements detect changes in steam pressure, oil and air flow. If steam demand increases the steam pressure will drop, a signal goes to the steam pressure controller (master) which is compared with the Set Value (sv)and the steam pressure controller output signal changes. This new output signal alters the set values of the oil and air flow controllers (slaves and in cascade) which in turn have their output signals altered to increase fuel and air supply. However, some of the items omitted in the foregoing are 1. Fuel supply increase must be preceded by an air supply increase otherwise bad combustion and contravention of clean air act could result. The converse is true for a rising steam pressure, an air supply decrease must be preceded by a fuel supply decrease. 2. Air-Fuel ratio must be controlled. To cover these two essentials a limiting relay and a Fuel-Air ratio controller are incorporated into the control system.

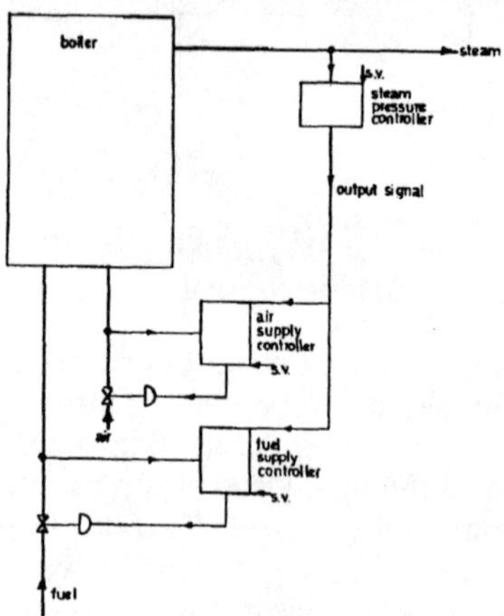

Figure 90
Steam pressure control

CARGO CONTROL SYSTEM

In Fig. 91 some of the main parts of a simplified cargo control system for a tanker are shown. A sequence control module suitably programmed would be situated in the control room, this is an essential feature as incorrect sequence of loading or discharging the tanks could lead to wide variations in trim. The planned programme can be interrupted by the operator changing over the hand/auto control, this may be necessary in emergency conditions.

If we consider loading the tank then both switches will transmit signals to the control module indicating that the tank is empty. Indication would also be given that the valve is in the closed position. Assuming the start signal is received by the sequence control module from the programme, then the cargo valve will be opened automatically and indication will be given by a feed back signal that the valve is open. When the cargo reaches the upper switch a signal will be given to close the valve and simultaneously a signal will be sent to another controller to open a valve in another tank. Indication that the cargo valve is fully closed would be fed back, however if the valve fails to close, alarm will be given, the operator can switch to manual and open valves in other tanks to avoid spillage.

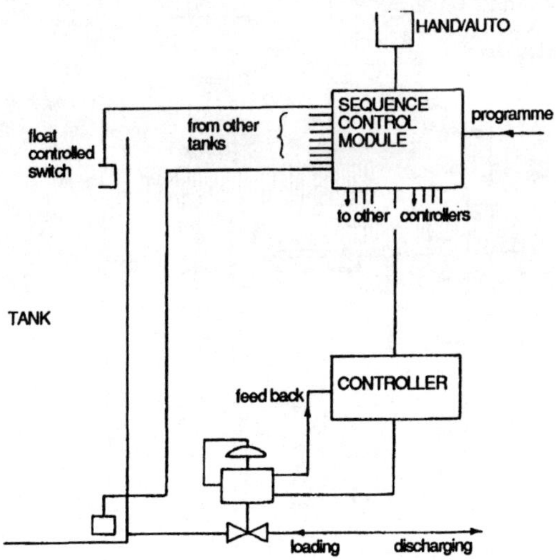

Figure 91
Cargo control system

Two independent level sensing systems would be installed, one of which would possibly be a float-controlled switch that would operate in the event of the first level detecting device failing. It would give alarm and re-activate the control system to the valve.

When discharging cargo, pumps would be automatically started, failure to do so would be indicated. Shoreside reception capability must be taken into consideration and programmed for. Whilst the main discharge programme is still in operation a stripping sequence would be automatically started and provision would be incorporated to enable the programme to be interrupted in any event.

For the control of cargo valve position on a tanker a hydraulic system may be used. A hydraulic power pack situated outside of the control room would consist of an oil reservoir, duplicated electrically driven submerged pumps, accumulators and emergency hand pump. In the control room would be remote controls and indicating units for the hydraulic system – pressure, level, temperature, flow, valve position and alarms.

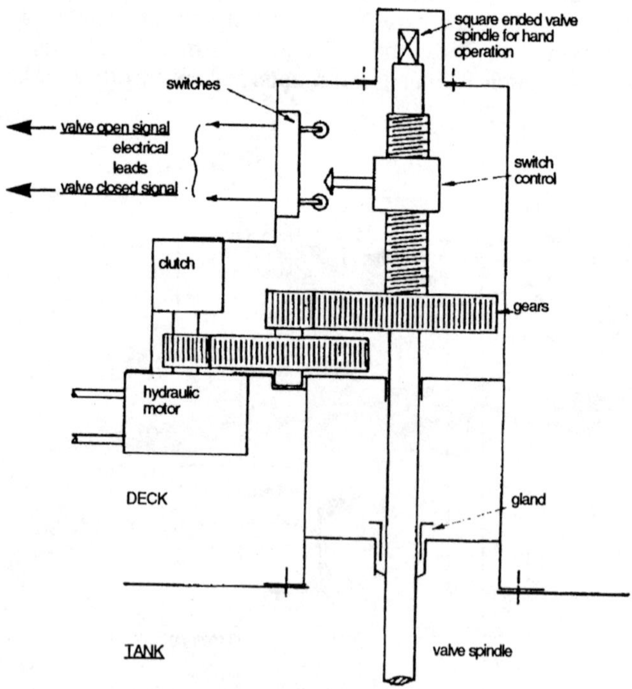

Figure 92
Rotary hydraulic actuator for a cargo valve

Fig. 92 shows diagrammatically a valve positioner of the hydraulic type. The hydraulic motor is totally sealed and drives the valve spindle through gearing. Incorporated with the positioner but not shown on the diagram is a cam operated three position sequence valve, the three positions of the valve give full hydraulic flow for the intermediate stage between opening and closing the valve, controlled torque and speed at point of closing and complete hydraulic cut off at the end of the opening stage. Intrinsically safe (circuits that are safe under normal and fault conditions) electrical leads, connections and switches transmit signals to the operator in the control room to give indication of valve position (such an arrangement can be incorporated with the pneumatic types and a local position indicator can also be incorporated with the hydraulic). In the event of failure of the hydraulic system, local manual operation of the valve can be achieved by disengaging the motor clutch and fitting a handwheel to the square ended valve spindle.

HYDRAULICALLY OPERATED HATCH COVERS. Two hydraulic rams actuated by an external power source and controlled locally at a control panel on the side of the hatch coaming are used to turn one pair of hatch covers, at their hinges, through 180°. Each pair of hatch covers are linked together by hinges and when the covers are opened by rolling back, they are moved into the vertical position, as described, automatically.

Reference has been made to control rooms in the foregoing control systems and the modern control room ergonomically designed, temperature and humidity controlled, contains a multifarious array of dials, switches, mimic diagrams, loggers, etc. Some of these items will now be examined.

DATA LOGGER

The term data logger is loosely used nowadays to describe a broad range of electronic systems that automatically collect and process data, some control and supervise the plant hence some data loggers would better be described as on-line computers.

Fig. 93 shows a very simple data logging system, the elements indicated perform the following functions:

1. Sensor-transducers. These are detecting conditions and changes in the plant under control such as pressure, temperature, flow, level, speed, power, position, and are converting the signals received into proportional d.c. electric outputs. The term transducer means conversion from one form

of signal to another, such as pressure to electrical output, or temperature to pressure output, and so on, but in the case of the data logger system generally the output signals are electric.

2. Scanner. This receives the d.c. outputs from the sensor-transducers, which are analogues of the physical functions being measured, into its channels. There may be from 10 to 1,000 different channels dealt with in rotation generally, but it can be equipped with random access facility.
3. Analogue-digital converter. This would incorporate the amplifier, the voltage signal received would be amplified and converted to frequency so that the signal is now in a suitable form for digital measurement and display.

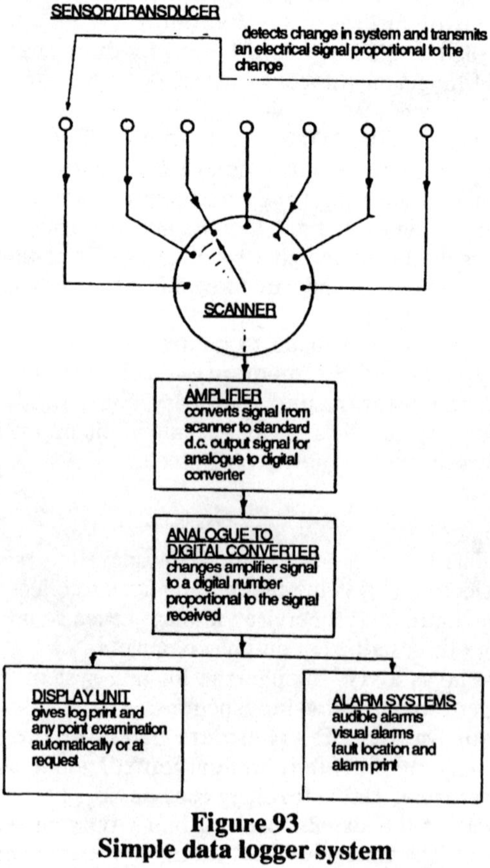

Figure 93
Simple data logger system

An on-line computer would have a programme stored within its memory and it would receive the digital signals from the analogue to digital converter and compare these with its programme, if these disagree then the computer would instigate corrective action. An automatic watch-keeping system for the engine room would have in addition to the foregoing, a shut down protection of the main engine with an over-riding facility in the event of unsafe manœuvring conditions, automatic change over to stand-by pumps and emergency generator, automatic bilge pumping and trend monitoring, this gives early indication of possible malfunctions.

In the event of an alarm limit being exceeded (limits are usually set by pins in a matrix board, but with trend monitoring, alarm limits can be varied automatically with ambient conditions) a klaxon will sound and a flashing window will identify the channel. The klaxon can be silenced by pressing an alarm acknowledged button, but the window will remain illuminated until the fault is remedied. The alarm condition and time of clearance will be automatically recorded.

All recorded data has time of recording associated with it and means are provided to enable the time to be adjusted to ships time, alarm print-out is often in red, normal in black. Most modern data logger systems have digital display of measured values whereas older types gave analogue display, such as voltmeters whose pointers traversed scales marked off in temperature, pressure, level, etc. Digital display is clearer, more accurate and reliable and by using frequency as the analogue of voltage there is greater freedom from drift.

Solid state devices and printed circuits are extensively used in the data logger system, these increase the reliability, simplify maintenance, are most robust, withstand vibration better and are cheaper to produce.

MIMIC DIAGRAM AND LEGEND

In a control room a mimic diagram of the engine layout indicating sensing points, as for example as partly shown in Fig. 94, together with a legend which enables the operator to identify the sensing points would be found. The operator can select the points for display of data *e.g.* pressure, temperature, flow, level etc., for display of individual data. Each of the points is sensed automatically in turn and if they are reading abnormal an alarm is given.

ANALOGUE AND DIGITAL DISPLAY. If we considered a pressure gauge of the type described in Chapter 1. The position of the pointer indicates the pressure of the fluid on the dial. This indication is an analogue of the pressure, something analogous to something else.

If we take the pressure being sensed and pass this signal to an analogue to digital converter (A/D converter) the output signal, usually electrical,can be used to actuate a digital display. *e.g.* 10.5 bar. A simple example of a digital display is the revolution counter on the Main Engine.

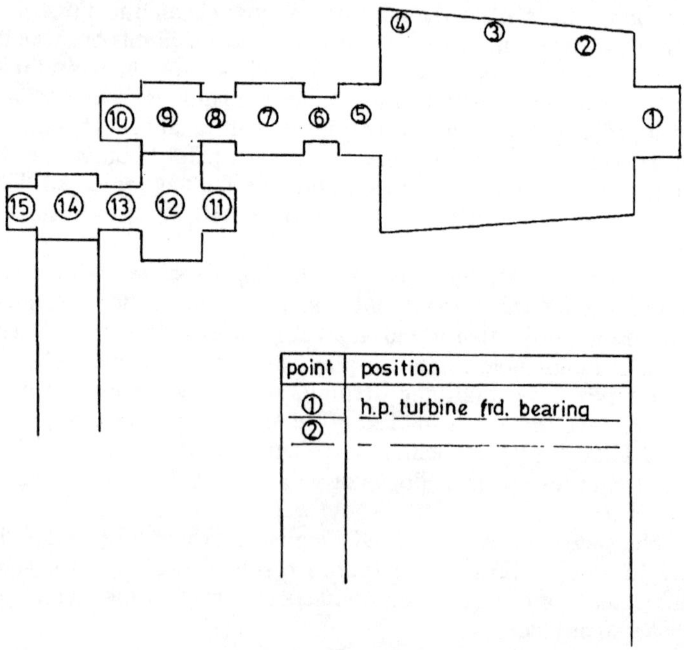

Figure 94

CARGO LEVEL INDICATION SYSTEMS FOR TANKERS

During the loading and discharging of cargo on a tanker it is essential that the controller is given immediate and accurate information of the level of cargo in the tanks whenever it is required, together with warning of approaching dangerous conditions in order that he may instigate appropriate corrective action to avoid spillage.

One system that supplies this information and at the same time is safe, reliable and simple is the pneumatic system. In order to give indication of level, air under pressure drives oil out of the pipe in the tank, the air pressure reached in the pipe must be a function of the head of oil in the tank. The air pressure is then transmitted to a pressure gauge, usually of the well manometer type, which indicates level and quantity.

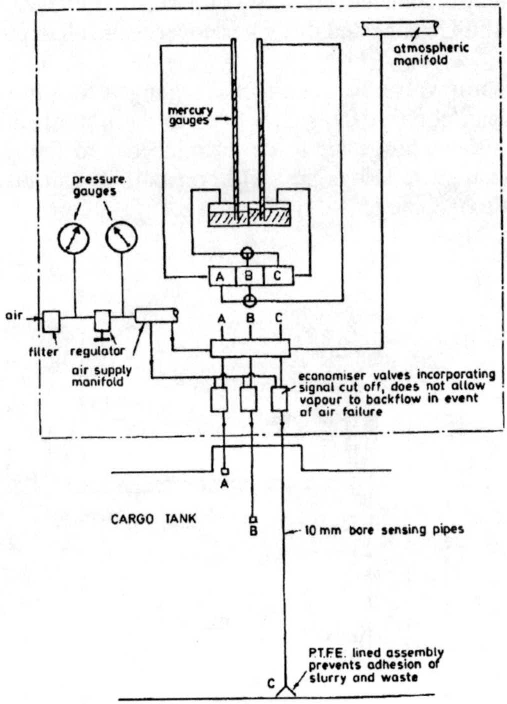

Figure 95
Pneumatic level detector

Fig. 95 shows a system that may be found on a modern tanker. A console unit situated in a control room contains the items outlined in the diagram. It is supplied with clean dry air and may have fitted an atmospheric manifold if the control room is pressurised.

Each mercury well type of manometer gauge has a zero adjust control that is used to give correction for temperature changes.

Operation is as follows: After correction for temperature change the gauge function selector switches would be set, they give the following functions.

Left hand column:
With A and B connected this gives last 2m of ullage.
With A and C connected this gives the first 2m of depth.
Right hand column:
With B and C connected this gives up to 2m of ullage.
With A and C connected this gives cover over full depth.

The selector valve can connect the gauges to atmosphere for zeroing, it can connect the signal lines to the h.p. manifold which allows a flush of purge air to be introduced into the signal lines whilst the gauge readings are held constant. It can also connect signal lines to the gauge.

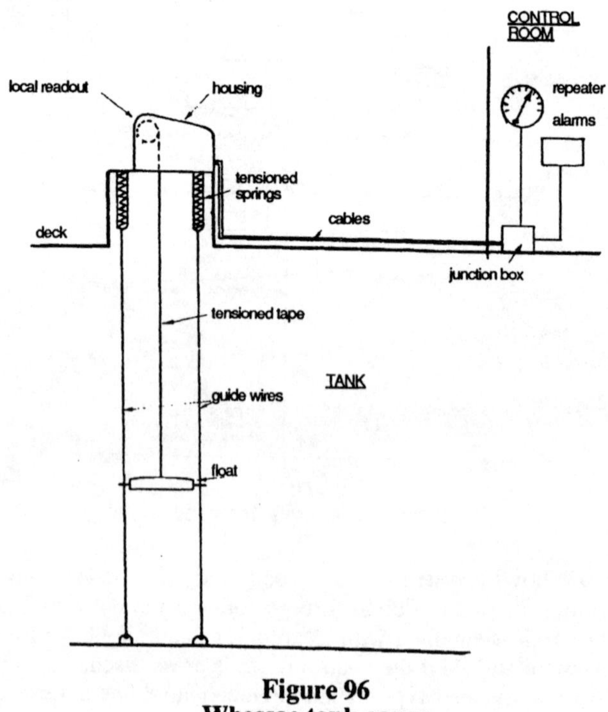

Figure 96
Whessoe tank gauge

WHESSOE TANK GAUGE

Used during loading, unloading and ballasting operations to provide automatic and continuous readout on a dial gauge or digital counter. They can be arranged to operate with standby batteries in the event of power failure and the transmission system is intrinsically safe.

The gauge measures liquid level by mechanical means, it consists of a stainless steel tape, with a float at one end, which is held at a constant tension by means of a tensator spring. A float runs on stainless steel guide wires and as it moves the tape winds or unwinds on its drum in the gauge housing. Rotation of the drum spindle is used to drive an a.c. transmitter motor which is connected to an a.c. synchronous repeater in the control room whose dial would be marked off in depth units.

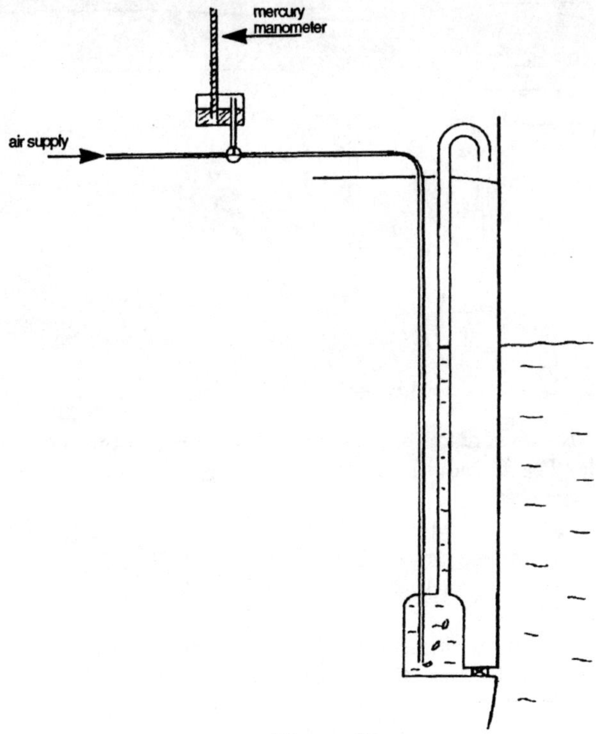

Figure 97
Draught indicator

Within the housing there is in addition to the a.c. motor a tape storage drum, local ullage indication, cam operated micro switches for level alarms, and when the gauge is not in use the float is hand wound up under the housing and locked in place.

Important items of information to the officer in charge of the loading or discharging operation is the draught of a vessel and its trim. For the determination at any instant of the draught of a vessel the pneumatic type of system is probably the most popular. Fig. 97 shows diagrammatically such a system.

FLOW RATE MEASUREMENT

There are many different methods of flow rate sensing, one of the simplest is the Venturi.

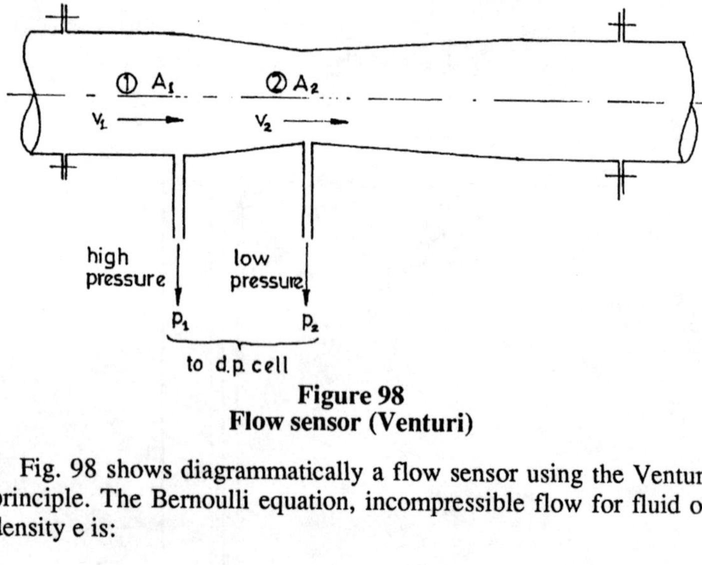

Figure 98
Flow sensor (Venturi)

Fig. 98 shows diagrammatically a flow sensor using the Venturi principle. The Bernoulli equation, incompressible flow for fluid of density e is:

$$\text{KE at 1} + \text{PE at 1} = \text{KE at 2} + \text{PE at 2}$$

$$\tfrac{1}{2} V_1^2 + P^1/e = \tfrac{1}{2} V_2^2 + P^2/e \quad \text{..........(a)}$$

Where KE is Kinetic and PE potential energy. This also assumes unit mass, negligible friction and shock losses. The continuity equation is:

$$V_1 A_1 = V_2 A_2 \quad \text{..........(b)}$$

By substituting for V_2 from (b) in (a) and using mass flow rate $\dot{m}$ as equal to eV_1A_1, then:

$$\dot{m} = k\sqrt{p}$$

Where p is the pressure difference ($p_1 - p_2$) and k is a meter constant. The pressure tappings would be fed into a differential pressure cell whose output would be arranged to be proportional to the mass flow rate $\dot{m}$ and would be used for display and /or control or both. The meter would be fitted directly into a pipe line or in a by-pass line if the main line is of large diameter.

TEMPERATURE SENSOR-TRANSDUCERS

Various types of temperature-sensor-transducers are used, choice depends upon many factors. High or low temperature measurement, environment, speed of response and cost.

THERMOCOUPLE. In 1821 Seebeck discovered that if a circuit consisting of two dissimilar metals had the junctions of the metals maintained at a different temperature, a current flowed in the circuit. The two metals used in the circuit form a thermocouple.

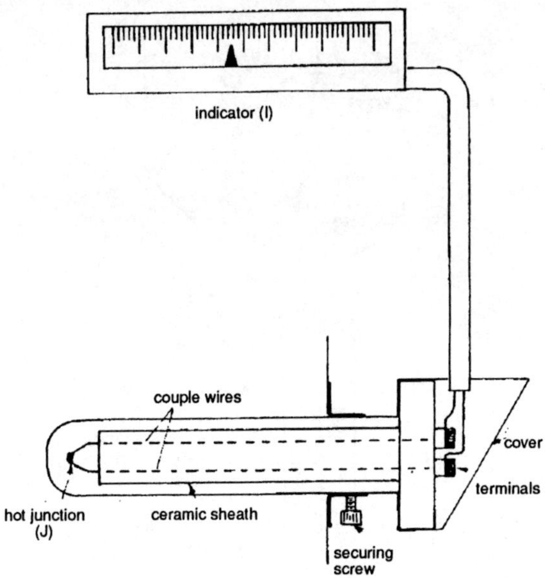

Figure 99
Thermocouple

A commercial type of thermocouple is shown in Fig. 99. The indicator (I) which is remote from the source whose temperature has to be measured forms one junction and together with the other junction (J) at a different temperature forms a thermocouple. The greater the temperature difference between I and J the greater is the e.m.f. (in millivolts) that will be produced.

Advantages are: simplicity, compactness, no external power required, quick to respond to changes of temperature, it is a relatively cheap telemetering device. It is used extensively for exhaust gas temperature, steam temperature, lubricating oil and cargo temperature, etc. Its main disadvantages are small changes in potential for large changes in temperature, needs good insulation, signals would require amplification if they were to be used in a control system.

Temperature sensing can also be achieved by using variation of electrical resistance with variation of temperature. The temperature

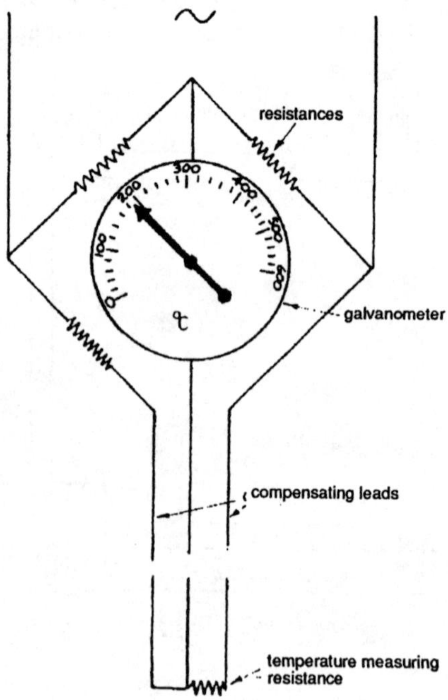

Figure 100
Resistance thermometer

sensing resistance would be incorporated into a Wheatstone bridge circuit whose balance would be upset if the temperature being measured varied.

In most metals, electrical resistance increases with temperature increase, but the semi-conducting thermistor has its resistance markedly decreased with temperature increase. Hence the thermistor, because of the big change in bridge potential for small change in temperature, is becoming increasingly popular as a temperature sensing device.

Fig. 100 shows diagrammatically such an arrangement, the leads to the temperature sensing resistance could be of considerable length and could pass through compartments at various temperatures. It is therefore important that they do not in any way effect the measurement of temperature so they are arranged to compensate each other.

Note: the term pyrometer is often used in conjunction with temperature measuring instruments. Generally, it is a term used for temperature measuring instruments operating above 500°C, the term thermometer being used for below 500°C. Hence any of the foregoing temperature measuring instruments could be classified as a pyrometer (pyro- Greek, fire).

WATER TIGHT DOORS

These may be vertically or horizontally operated in slides, usually the latter, by a remote controlled hydraulic actuator. Local hand control, mechanical (hand pump) and hydraulic is provided. The remote control and power unit is situated above the bulkhead deck this also incorporates a hand operated pump for door operation in event of power supply failure. The system should be checked regularly for leaks and tested by opening and closing the doors during fire drill. Tracks must be kept clean and free of obstructions to ensure smooth operation.

FIRE DETECTION METHODS

FIRE PATROLS. These are not normally carried out on a regular basis upon most vessels but they should be conducted (1) immediately prior to, or upon sailing. A thorough inspection of the vessel being made especially in hold compartments, stores, engine and boiler rooms, etc. (2) when the vessel has been vacated by shipyard personnel whilst the vessel is in port undergoing repair. Someone may have been using oxy-acetylene burning or welding equipment on one side of a bulkhead totally unaware that the beginnings of a fire were being created on the other side of the bulkhead.

The patrol should, in addition to looking for fire, assess and correct any possible dangerous situation, *e.g.* loose oil or paint drums, incorrectly stored chemicals, etc.

FIRE ALARM CIRCUITS. These consist of an alarm panel, situated outside of the machinery spaces, which gives indication of the fire zone. Zone circuits, audible alarms and auxiliary power supply (Fig. 101).

Circuits. When the contacts in a detector head close (open under normal conditions) they short the circuit and cause operation of the audible fire alarm. The lines in the circuit are continuously monitored through 1 to 2 and 3 to 4, hence any fault which develops, *e.g.* damaged insulation, break in the cable, causes the system failure alarm to sound.

Power Failure. In the event of failure of mains supply power, automatic auxiliary power is supplied from fully charged stand-by batteries for up to 6 hours. Most systems operate on 24V d.c., however, for those operating at mains supply of 220V a.c.an inverter converts the 24V d.c. to 220V a.c.

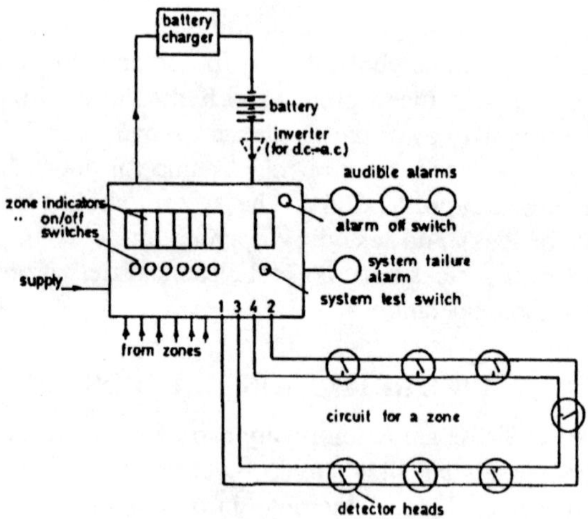

Figure 101
Fire alarm circuit

Audible Alarms. The fire alarm is usually an intermittent audible signal whereas fault and manual test are normally a continuous audible signal.

FIRE DETECTOR HEADS Various types are available for fitting into an alarm circuit, choice is dependent upon fire risk, position, area to be covered, volume and height of compartment, atmosphere in the space, etc. To economise and simplify, standard bases are generally used in the circuit into which different types of detectors can be fitted.

HEAT SENSORS These may be fixed temperature detectors, rate of rise detectors or a combination. Rate of rise detectors do not respond and give alarm if the temperature gradually increases, *e.g.* moving into tropical regions or heating switched on.

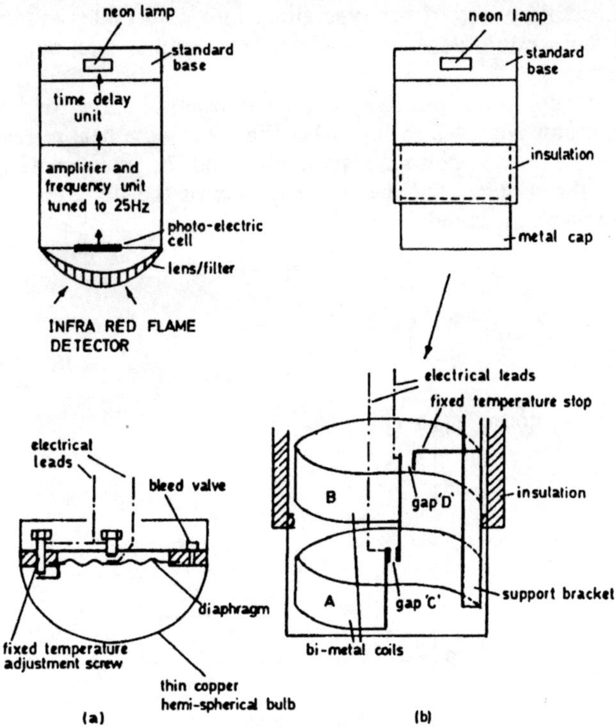

Figure 102
Fire detectors

Fig. 102(a) *Pneumatic Type.* Increase in temperature increases the air pressure inside the hemi-spherical bulb, if the bleed of air through the two way bleed valve from the inside of the bulb is sufficient the diaphragm will not move up and close the contacts. If however the rate of rise of temperature causes sufficient pressure to build up inside the bulb to close the contacts, alarm will be given. In either case a bi-metal unit will at a pre-determined temperature close the contacts on to the fixed temperature adjustment screw, giving alarm.

Fig. 102(b) *Bi-metal Coil Type.* Two bi-metal coils attached to a vertical support bracket are encased in a protective metal cap. When the temperature increases A will move to close the gap C at a faster rate than B moves to maintain the gap, this is due to B being better insulated from the heat than A. If the rate of rise of temperature is sufficient, gap C will be closed and alarm given. At a *fixed* temperature gap D, then gap C will be closed, giving alarm.

Quartzoid bulbs of the type fitted into a sprinkler system are fixed temperature detectors used for spaces other than engine and boiler rooms.

Relative Points. Sensitivity: a typical response curve for a rate of rise detector is shown in Fig. 103. The greater the heat release rate from the fire the poorer the ventilation and the more confined the space, the quicker will be the response of the detector and the sooner an alarm sounds.

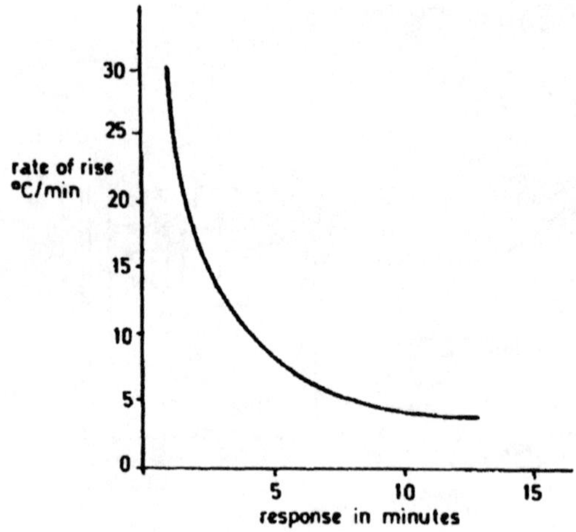

Figure 103
Detector response curve

Fixed temperature setting depends upon whether the detector is in accommodation or machinery spaces and can vary from 55°C to 70°C.

The detector is useful for dusty atmospheres as it is completely sealed but it does not give as early a warning of fire as other types of detectors. It can be tested by a portable electric hot air blower of muff.

INFRA RED FLAME DETECTOR. Fig. 102 also shows in simplified form this type of flame detector. Flame has a characteristic flicker frequency of about 25Hz and use is made of this fact to trigger an alarm. Flickering radiation from flames reaches the detector lens/filter unit, which only allows infra-red rays to pass and be focused upon the cell. The signal from the cell goes into the selective amplifier, which is tuned to 25Hz, then into a time delay unit (to minimise incidence of false alarms, fire has to be present for a pre-determined period), trigger and alarm circuits.

Relevant Points. Very early warning of fire is possible, suitable for areas where fire risk is high, *i.e.* machinery spaces – but not in boiler rooms where naked flame torches are to be used for igniting oil. Reflected radiation can be a problem in boiler rooms and from running machinery. Obscuration by smoke renders it inoperative. It can be tested by means of a naked flame.

PHOTO-ELECTRIC CELL SMOKE DETECTORS. Three types are in use, those that operate by light scatter, those that operate by light obscuration and a type which combines scatter and obscuration.

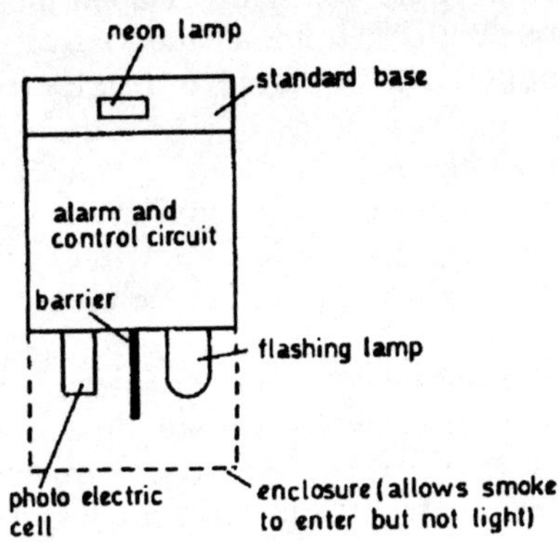

Figure 104
Smoke detector. Light scatter type

Light Scatter Type (Fig. 104). A photo-cell separated by a barrier from a semi-conductor intermittently flashing light source are housed in an enclosure whose containment allows smoke but not light inside. When smoke is present in the container, light is scattered around the barrier on to the photo-cell and an alarm is triggered.

Relevant Points. Smoke may be present without much heat or any flame, hence this detector could give early warning of fire. Photo-cells and light sources are vulnerable to vibration and dirt. Testing can be done with smoke from a cigarette.

The light obscuration type is used in oil mist detectors for diesel engine crank cases and the obscuration/scatter type is to be found in the detecting cabinet of the carbon dioxide flooding system.

STANDARD BASES. The standard bases shown in the figures for the various detector heads have a neon light incorporated which flashes to indicate which detector head has operated. Detector heads can be simply unplugged from the base and tested in a portable test unit which has an adjustable time delay, audible alarm and battery.

COMBUSTION GAS DETECTOR. A circuit diagram of a combustion gas detector is shown in Fig. 105. Two ionisation chambers connected in series contain some radioactive material which emits a continuous supply of ionising particles.

The detecting chamber is open, the reference chamber closed and operating at a constant current since it contains air which is being ionised and the applied potential ensures that saturation point is passed. Current strength is dependent upon the applied potential, since if the potential is low not all the ionised particles reach the electrodes, some will combine with electrons and thus be neutralised.

When the potential reaches a certain value all the ions formed reach the electrodes giving saturation. Beyond this, the current will remain approximately constant irrespective of any further increase in potential. In this way the reference chamber has a constant resistance.

If combustion particles, visible or invisible, pass through the open detecting chamber the current will drop since the combustion products are made of larger and heavier particles than normal gas molecules. When ionised, the particles are less mobile than ionised air particles and because of increased bulk and lack of mobility, can readily combine with particles of opposite charge and hence be neutralised. The effect is to greatly increase the resistance of the detecting chamber, this change in resistance produces a substantial change in the potential at the centre point B.

Normal voltage A to C is 220 volts, A to B 130 volts, B to C 90 volts. When voltage shift, due to increasing resistance in the detecting chamber, reaches 11OV across BC this is sufficient to trigger a discharge in the valve from 2 to 3, the capacitor then unloads itself across 2 to 3 encouraging a discharge from 1 to 3, bypassing the chambers and causing heavy current flow through the alarm relay and the alarm to sound.

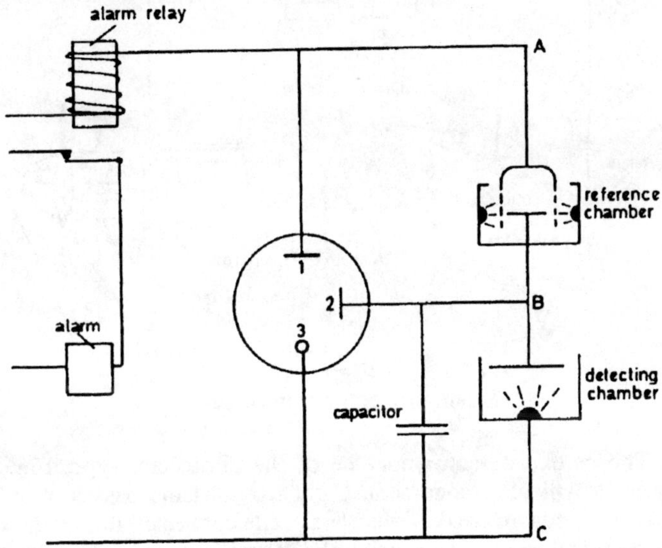

Figure 105
Combustion gas detector

It can be tested by cigarette smoke or the use of butane gas delivered from an aerosol container. It is a very sensitive fire alarm and a time delay circuit may be incorporated to minimise the incidence of false alarms.

A smoke detection alarm and CO_2 flooding arrangement is frequently used for hold compartments. Small diameter sampling pipes are led from the holds to a cabinet situated on the bridge. Air is drawn continuously through these pipes to the cabinet by suction fans, which deliver the air through a diverting valve into the wheelhouse.

When a fire breaks out, smoke issues from the diverting valve into the wheelhouse, warning bridge personnel of the outbreak. Simultaneously, a smoke detector in the cabinet sets off audible alarms, hence if the bridge is unoccupied notice of outbreak of fire is still obtained.

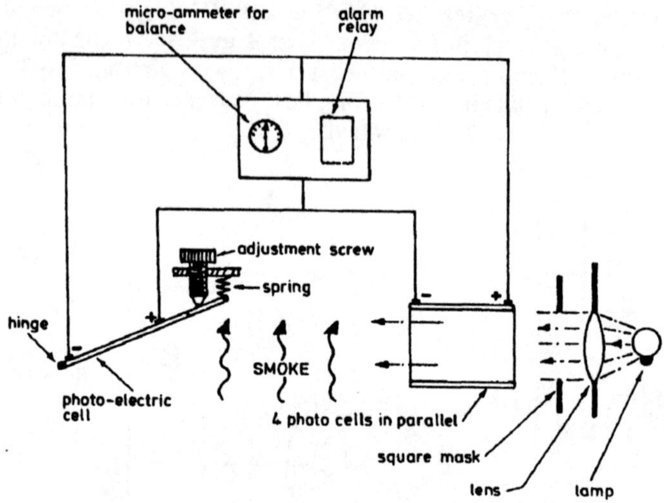

Figure 106
Smoke detector (photo-cell type)

The smoke detector may be of the photo-cell type, Fig. 106 shows a typical arrangement. Light from a lamp passes through a lens and a square mask. Four photo-cells connected in parallel form an open ended box from which the square beam of light is emitted. The light falls onto another photo-cell, this cell which can be manually tilted is adjusted to a position such that its electrical output balances that of the four cells in parallel. When smoke passes through, the amount of light falling on the single cell is reduced, and due to reflection the light falling on the four cell block is increased. Current then passes through the circuit and a relay will operate audible alarms on the bridge, in the engine room and in the accommodation.

Inspection of the dark chamber in the cabinet will enable the affected compartment to be identified as smoke will be seen issuing from a labelled illuminated chimney. Then by opening up the requisite number of CO_2 bottles for the compartment and operating the appropriate change over valve in the bottom of the cabinet, CO_2 flooding of the affected compartment can be achieved.

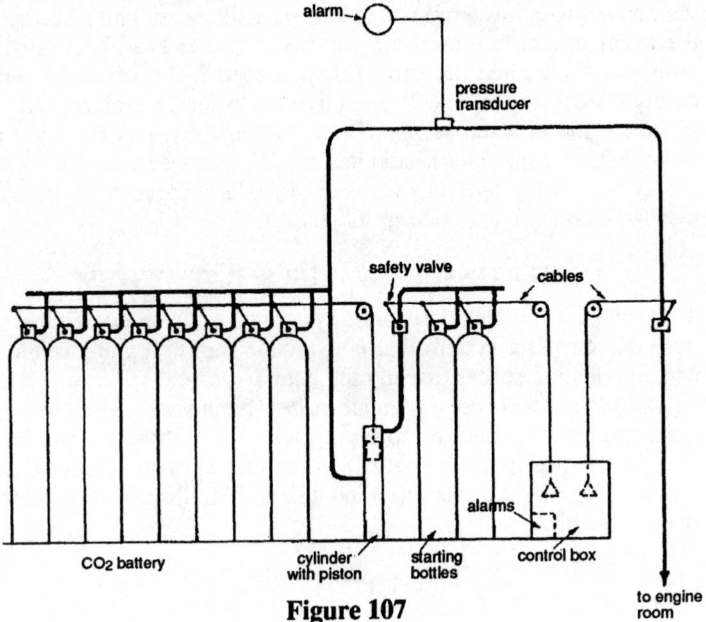

Figure 107
CO2 total flooding system

Some engine rooms are fitted with a CO_2 total flooding system. To operate, first ensure that the compartment is evacuated of personnel and sealed off. This necessitates closing all doors to the engine room, shutting down skylights, closing dampers in vents and stopping ventilation fans. Pumps should also be stopped and collapsible bridge valves closed. In a modern vessel the sealing off can be done by remote control from the fire control station generally using a compressed air or hydraulic system.

The door of a steel control box situated at the fire control station would then be opened, this operates a switch which may have a dual purpose. One is to operate audible and visual alarms in the engine room spaces, the other may be to shut off ventilation fans. The CO_2 direction valve handle inside the box would then be pulled and this would be followed by the gas release. This would give 40% saturation of CO_2 in the compartment of which 80% enters within two minutes.

Another type of fire detection, alarm and extinguishing system is the well known sprinkler system. Briefly it consists of an air pressurised water tank from which a tree of pipes spreads through the accommodation terminating in heat sensing sprinkler heads. In the event of fire in a compartment the sprinkler head heat sensing bulb will shatter and the valve it kept closed will open and a spray of high pressure water will be delivered to the fire, alarm will be given for the affected section. Each section (in which there are no more than 150 sprinkler heads) has its own alarm system that can be easily tested by opening the section drain valve, this gives a discharge equivalent to one sprinkler head.

HIGH PRESSURE WATER SPRAY SYSTEM

This can be a completely separate system or it can be interconnected with the sprinkler system that is available for fire extinguishing in accommodation spaces (usually the latter).

The system incorporates an air vessel, fresh water pump and salt water pump all connected to piping which is led to sections, each section having its own shut-off valve and sprayer heads, which unlike the sprinker system have no quartzoid bulbs or valves but are open (Fig. 108).

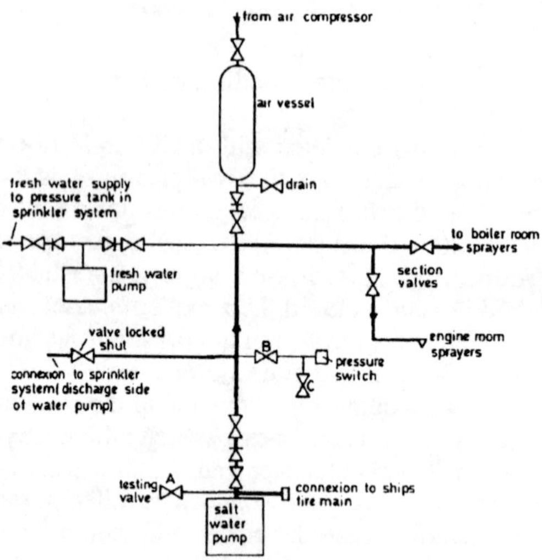

Figure 108
High pressure water spray system

With all section valves closed the system is full of fresh water under pressure from the compressed air in the air vessel. When a section valve is opened water will be discharged immediately from the open sprayer heads in that section, pressure drop in the system automatically starts the salt water pump which will continue to deliver water to the sprayers until the section valve is closed.

After use the system should be flushed out and recharged with clean fresh water.

The air vessel is incorporated into the system to prevent the pump cutting in if there is a slight leakage of water from the system.

To test: this should be carried out at weekly intervals, open A, close B, open C; the pump should automatically start and discharge from A. This avoids having to refill the system with fresh water.

INERT GAS GENERATOR

This was originally developed to supplement $C0_2$ flooding systems. Since, if a fire occurred on board a ship at sea and the fire was extinguished through using all the CO_2 available and a further outbreak of fire occurred, the situation could be dangerous.

In a compartment wherein there is an outbreak of fire, the minimum percentage of oxygen in the atmosphere in the compartment which will allow combustion to proceed varies with different materials between 12 to 16 per cent approximately. Hence if the oxygen content of the compartment can be reduced below 12 per cent, insufficient would be present to allow combustion to continue. This reduction in oxygen content can be achieved by employing a generator which will supply inert gas which is heavier than air, so displacing the atmosphere in the compartment.

The generator consists of a horizontally arranged brick lined furnace cylindrically shaped and surrounded by a water jacket. This is connected to a vertical combustion chamber in which water spray units and Lessing rings (cylinders of galvanised metal arranged to baffle the gas flow) are fitted. A water cooled diesel engine, usually fitted alongside the generator, drives a fuel pump, a constant volume air blower and an electric generator. The electric generator supplies current to an electric motor which in turn drives the cooling water pump, motor and pump are usually situated at the forward end of the shaft tunnel. By fitting the cooling water pump in the shaft tunnel and having it connected to the wash deck line, this pump can also be used as an emergency fire pump. Cooling water for the gas generator can also be supplied by ballast and general service pumps in the engine room, the amount of water required is approximately

545 litres per hour for every 27.7 m³ of inert gas produced.

The oil fuel burner is initially lighted by means of high tension electrodes, the electrical supply being through a small transformer. A constant pressure regulator is fitted to the oil supply line to the burner along with a control valve.

A control panel for the gas generator incorporates a CO_2 recorder, water and oil fuel alarms and pressure gauges. In the gas piping system leading from the combustion chamber, condensate traps and drains are fitted.

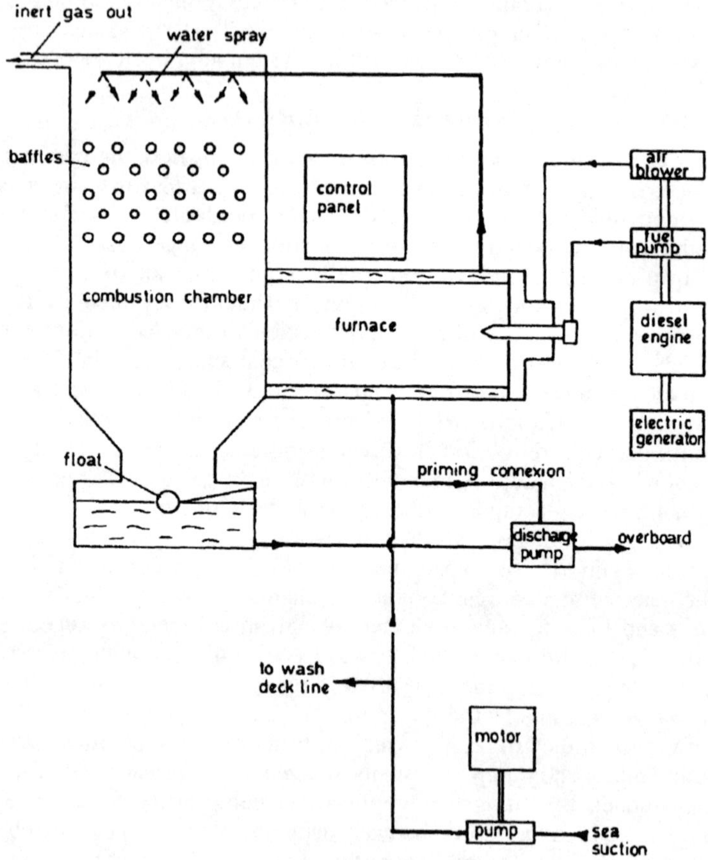

Figure 109
Inert gas generator

The following is an approximate analysis of the gas generated.

Oxygen	0-1per cent
Carbon monoxide	Nil
Carbon dioxide	14-15 per cent
Nitrogen	85 per cent

Remainder, unburnt hydrocarbons and oxides of nitrogen.

FOAM COMPOUND INJECTION SYSTEM

Fig. 110 shows diagrammatically the compound injection system often found on tankers for deck and machinery spaces. Tank and pumps may be situated wherever it is convenient.

Foam compound is drawn from the sealed tank by the compound pump, air enters the tank through the atmospheric valve (this being linked to the compound valve). Both open simultaneously and delivery is to the automatically regulated injector unit. The injector unit controls the amount of water to compound ratio for a wide range of demand by the foam spreaders etc. A fire pump delivers the foam making solution at sufficient pressure to the deck monitors (multi-directional type foam guns) so that foam can reach any part of the deck, or to the spreaders for machinery spaces.

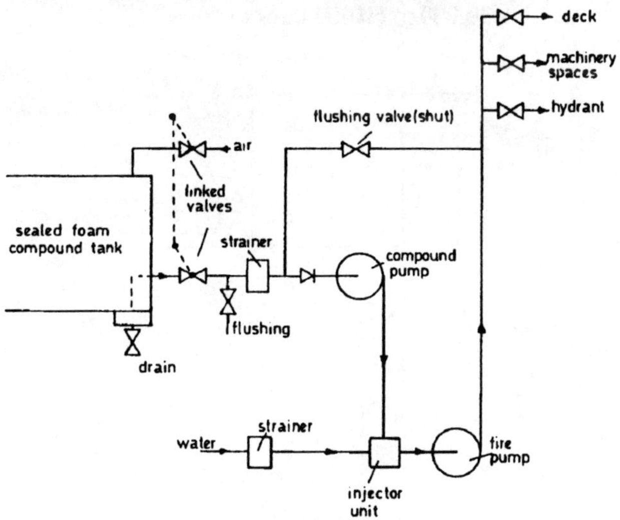

Figure 110
Foam compound injection system

To bring the system into operation it is only necessary to open the linked air/foam compound valves and start the pumps. After use the system must be thoroughly flushed through and recharged.

In chemical foam installations the principal disadvantage is the deterioration of the chemicals and chemical solutions, hence regular checking is necessary to ensure the system is at all times capable of effective operation. However, with the chemical foam system good quality uniform foam is capable of being produced.

With mechanical foam systems, storage and deterioration of the foam compound presents no difficulty, which is one of the reasons why this particular type of system is generally preferable.

HIGH EXPANSION FOAM SYSTEM

This recently introduced foam system has been recognised by the DTp as an alternative fire extinguishing medium for boiler and engine room compartments.

The generators are large scale bubble blowers which are connected by large section trunking to the compartments (Fig. 111)

A 1½m long, 1m square generator could produce about 150 m^3/min of foam which would completely fill the average engine room in about 15 minutes. One litre of synthetic detergent foam concentrate combines with 30 to 60 litres of water (supplied from the sea) to give 30,000 to 60,000 litres of foam.

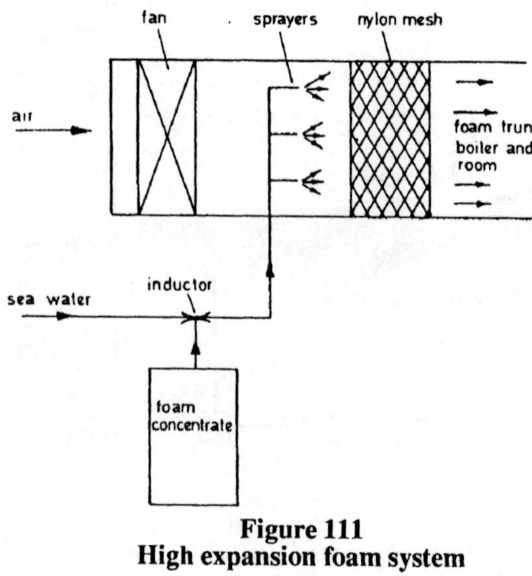

Figure 111
High expansion foam system

Advantages:
(I) Economic; (2) Can be rapidly produced; (3) Could be used with existing ventilation system; (4) Personnel can actually walk through the foam with little ill effect.
Disadvantages:
(1) Persistent, could take up to 48 hours to die down in an enclosed compartment; (2) Large trunking required; (3) Should be trunked to bottom of compartment to stop convection currents carrying it away.

OXYGEN ANALYSER

Fig. 112 shows in simplified form a fixed oxygen analyser. Two platinum wire resistances are heated by current from an a.c. bridge (Wheatstone), one of the resistances is in a magnetic field which attracts oxygen, as oxygen is paramagnetic. Gas enters the resistance chamber via a diffuser and convection currents are set up around the resistance in the magnetic field which cools this resistance relative to the other. The bridge is then unbalanced and the amount of unbalance is proportional to the oxygen content of the gas which is displayed on the galvanometer.

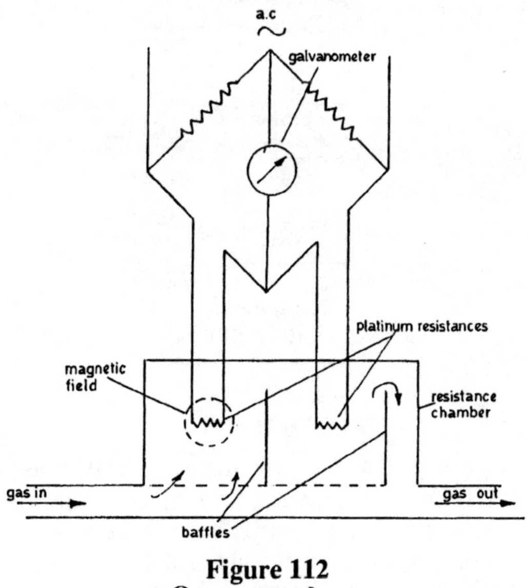

Figure 112
Oxygen analyser

COMBUSTIBLE GAS DETECTOR

A portable battery operated combustion gas detector is sometimes used to check atmospheres of tanks. The most commonly used is the catalytic filament gas indicator. This uses a heated platinum filament to catalyse the oxidation of combustible vapours. Vapour oxidation can occur even outside the limits of flammability and the heat generated at the filament raises its temperature and alters its electrical resistance. Resistance is approximately proportional to the concentration of vapour and as this resistance forms one limb of a Wheatstone Bridge network the galvanometer will display ,when the bridge is unbalanced, the condition of the gas sample drawn into the detector by the hand squeezed bulb. To avoid false readings the meter should be checked against test kits provided before use. Inaccuracy can be dangerous, the filament may need replacing.

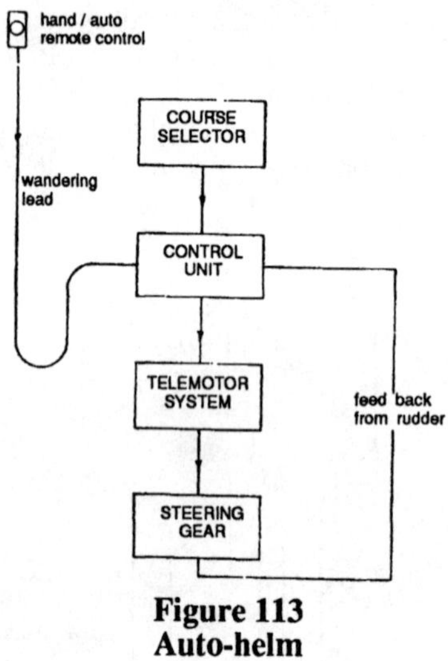

Figure 113
Auto-helm

AUTO-HELM

Fig. 113 shows in block diagram form the well known Kelvin Hughes solid state type of Autohelm. In operation the course selector receives directional information from the compass fitted, and if the vessel goes off course the Autohelm takes corrective action. A deviation of half a degree on maximum sensitivity causes the unit to function and restore balance. The appropriate amount of counter rudder is automatically applied to prevent oversteering in any state of trim, and additional controls are provided to determine the initial amount of rudder applied and to compensate for the yawing of the vessel.

Audible indication that the equipment is functioning correctly is given by high and low pitched notes which coincide with rudder movements to starboard and port respectively. Also, a sensitivity control slows down the application of rudder should it be found to be working too much.

A remote control hand held unit on 10 metres of wander cable enables an operator to control the system from the wings of the bridge, when the remote control is switched off the auto-helm will automatically bring the vessel on to the compass course set at the time.

ELECTRO-MAGNETIC LOG

This sensor-transducer unit gives a vessel's speed, ahead and astern, and distance covered. It operates on the electro-magnetic principle of potential difference generation by a moving conductor in a magnetic field.

A probe unit consisting of an electro-magnet and two electrodes is lowered by an electric motor through an opening in the ship's bottom. As the vessel moves through the water a potential difference proportional to the vessel's speed is generated by a moving conductor (the water) in the magnetic field provided by the electro-magnet, and this is picked up by the two electrodes.

An electric current in the range O to 20mA d.c. is produced for transmission to the speed repeaters and is electronically integrated to provide two pulse outputs with respect to time, one occurs every 10 metres for feeding into the radar and the other every 0.1 of a nautical mile is fed into the distance repeater.

The opening in the ship's bottom can be sealed off by a sea valve when the probe unit is raised (this allows for inspection and maintenance). Interlocks are provided to ensure that the probe cannot be lowered onto a valve not fully open.

Accuracy is about ±1% of maximum speed and the signal generated can be used in a data logger system if it is required.

SHIPS TELEGRAPH

Fig. 114 shows diagrammatically the Chadburn "Synchrostep"

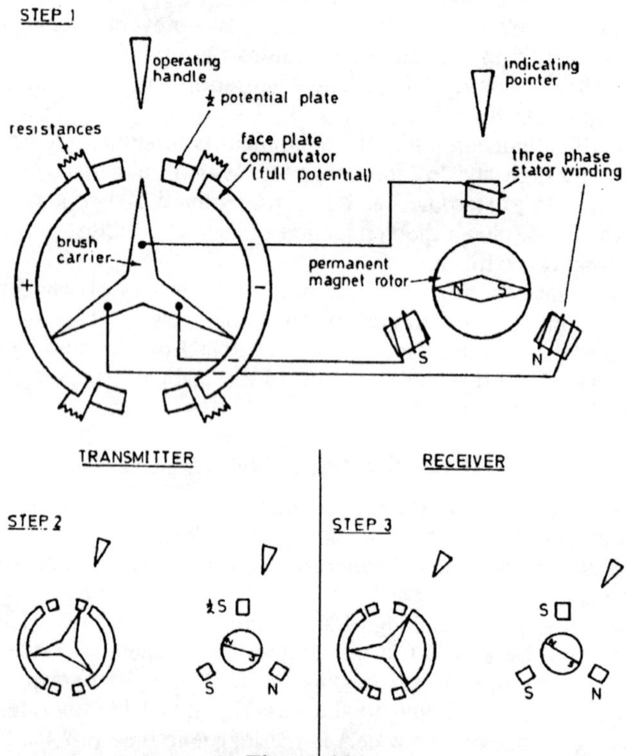

Figure 114
Chadburn Synchrostep electric telegraph

electric telegraph. The transmitter consists of a face plate commutator and brush carrier, the brush carrier being connected to the operating handle and indicating pointer or pointers (top and side).

Moving the operating handle rotates the brush carrier and this causes alteration in polarity in the windings of the receiver. A permanent magnet rotor (wound rotor for a.c.) will immediately take up a new position that corresponds exactly to the transmitter request, this rotor carries an indicating pointer.

Three positions are shown in the diagram with 15 degree steps between each position but more positions with 15 degree steps are used than shown, generally 12 indicative steps over 180 degrees.

The instruments are robust, watertight, contain night

illumination, alarms and can operate on a.c. or d.c. When the telegraph is operated the alarms will sound continuously until answered and in the event of current failure an alarm can be arranged to give visual and/or audible indication.

A telegraph order and recorder system can give a print out similar to that shown below.

	Day of Year	Time h m	Bridge Command	Engine Room Acknowledgement	Prop Shaft Revs.	Ahead + Astern -
Print-out line	200	10 40.5	4	4	100	+

The paper advances one print-out line every half minute interval. The recorder system has the advantage of freeing the operator from the recording chore, if necessary print out of all important parameters can be had on request.

STABILISERS

If a vessel is heeled by an external force and this force is suddenly removed the vessel will roll with a rolling period that is nearly constant. This period (*i.e.* time to roll from either port to starboard and return) is called the ship's "natural rolling period". When the waves that are applying the force which causes rolling have a period that synchronises with the "natural rolling period" resonance may occur (hence resonant rolling) and the amplitude of roll may reach dangerous proportions. The ships heading must be changed to alter the period at which the vessel is encountering the waves.

A common type of anti-roll device is a pair of stabilising fins which project in a horizontal plane from each side of the ship at approximately amidships about bilge level, and retractable into watertight compartments when not in use. Each fin, which is attached to a vertical shaft at its inboard end, is of aerofoil section and rectangular plan shape and is capable of being rotated about its polar axis up to an angle of 20 degrees above and 20 degrees below the horizontal.

When the ship is rolling and the stabilising fins are brought into operation, they are swung out from their fore and aft position in their water-tight compartment by electro-hydraulic action. The tilting movement is controlled by electrically driven gyroscope-units which anticipate the angular magnitude and velocity of roll, and a rotary-vane hydraulic unit housed in the fin hub imparts the

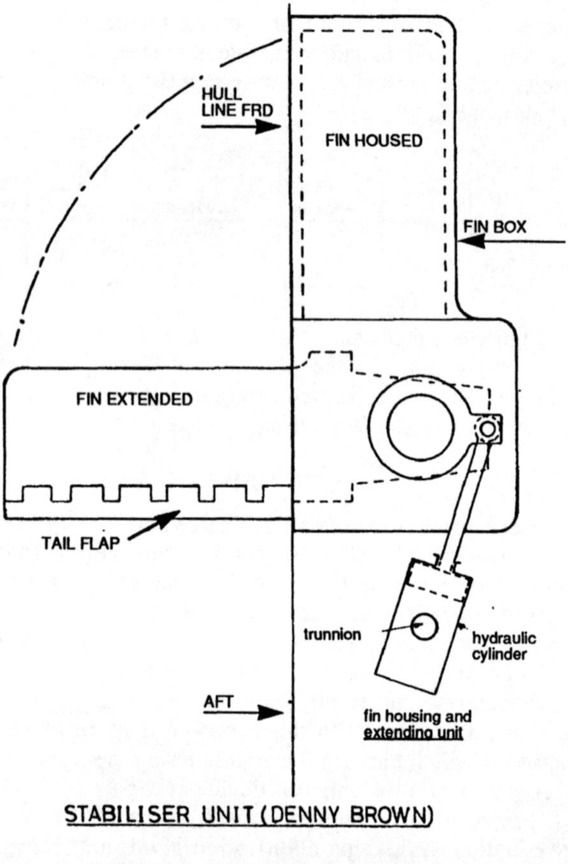

Figure 115
Stabiliser unit (Denny Brown)

necessary motion. The fins tilt in opposite directions to each other and their leading edges dig into the water stream due to the forward motion of the ship, in such a way that a downward resultant force acts on the fin on the ascending side of the ship, and an upward resultant force on the fin on the descending side, thus setting up a righting couple to counteract the rolling motion. As the direction of roll changes, the fins flick over to the opposite angle of tilt.

A tail-flap fitted on each fin moves in the same direction as the fin but through a greater angle, to increase the stabilising effect.

Control of the unit is exercised from the bridge, there are three main hand-operated switches: (i) Signal to order start up of motors. (ii) To extend or house fins. (iii) Stabilisers "on-off." Control panels carry indicator lamps to show the in or out positions of the fins and "Fin angle indicators" show fin movement during stabilising operation.

STABILISATION BY USING TANKS. Three systems are used: – Passive tanks,Controlled passive tanks and Active tanks.

1. Passive tanks.
   Fig.116 shows the arrangement. Two wing tanks are connected by a duct which incorporates a series of baffles. Water partly fills the tanks and moves across the system in the direction of roll. When the vessel starts to return after a roll the water slowed by the baffles and its inertia acts to produce a stabilising moment. The quantity of water in the tanks must be tuned to avoid dangerous rolling which could occur if water movement and ship movement synchronise.Water level therefore is adjusted according to the loaded condition of the ship.

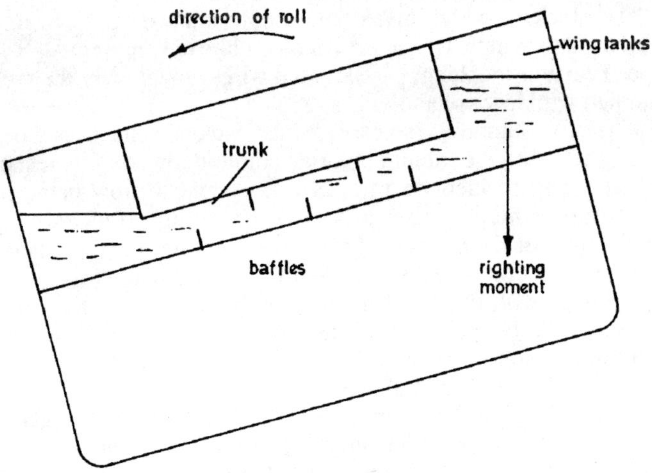

Figure 116
Passive tank system

2. Controlled passive tanks. This is similar to the previous system except that the water movement is controlled by valves which are positioned by a control system similar to that described for the fin stabilisers.
3. Active tanks. Water is driven by a constantly running uni-directional impeller through a series of controlled valves which are actuated through a gyroscope system similar to that used in the fin stabiliser. Movement of the water from wing tank to wing tank is in the opposite direction to the roll.

AIR CONDITIONING

Ambient temperature and relative humidity need to be kept within limits for comfort of passengers and personnel.

Relative humidity is the ratio of the mass of water vapour present in unit volume of air to the mass of water vapour required to saturate the same volume of air, usually expressed as a percentage.

To obtain relative humidity two temperatures are often used, one is the ambient temperature the other is the dew point of the air, *i.e.* the temperature at which the air would be saturated with water vapour and condensation would occur.

A commercial hydrometer (*i.e.* an instrument which measures moisture content) used for control and remote reading is the "Dewcel", manufactured by Foxboro-Yoxall. Briefly and simplified its description and operation are as follows:

The element, which gives dew point temperature, consists of a tube around which is wound lithium chloride impregnated glass tape. Two non-touching, parallel gold wires wound over the tape are supplied with low voltage a.c.

If relative humidity increases, water vapour condenses from the atmosphere into the lithium chloride solution lowering its resistance and allowing an increase in electrical current to flow between the two wires, which in turn produces a heating effect raising the temperature of the element. This action of water vapour absorption and heating continues until the lithium chloride solution is in equilibrium with the air. A resistance thermometer inside the tube measures the temperature of the element. This would be the dew point temperature after allowance has been made for using lithium chloride solution instead of water.

Use of the signal from the element, together with a similar form of signal for ambient temperature can be used in control equipment for air conditioning units, or to give local and or remote indication of dew point and ambient temperature together with relative humidity.

SELECTION OF EXAMINATION QUESTIONS DESCRIPTIVE

1. Sketch and describe a water-tube boiler which incorporates an economiser. State the purpose and advantages of the economiser.
2. State the meaning of the term "feed-back" and explain its function in a control system.
3. Describe step by step the procedure for lighting up an oil fired boiler and state the precautions that must be taken.
4. Sketch and describe a pneumercator gauge suitably equipped for remote reading of cargo in a tanker.
5. (a) Explain fully the meaning of the terms "indicated power" and "brake power" and state the relation between them.
   (b) Describe the method of determining the indicated power of a ship's main engine.
6. Sketch a type of thermocouple, give examples where they are used in marine instrumentation and state the advantages and disadvantages of a thermocouple over other temperature measuring devices.
7. Sketch and describe a regenerative condenser, explain its function and give reasons for the use of this type of condenser in turbine installations.
8. Describe and illustrate the cycle of operations in an opposed piston engine and explain the meaning of the expression "dual-combustion" in connection with this type of engine.
9. Sketch and describe a data logging system and discuss the merits and demerits of solid state devices in such a system.
10. Describe an automatic sprinkler system and explain how it can be tested during statutory fire drills.
11. What is "saturated steam" and "superheated steam"? State clearly the difference between the two and explain how superheated steam is generated. Why is superheated steam preferred in a turbine?
12. Explain how remote indication of valve position is achieved in a hydraulic valve control system.

13. What is an exhaust gas turbo-charger and where is it used? Make a diagrammatic sketch of a turbo-charger, describe how it works and state the advantages of turbo-charging.
14. Describe a hydraulic control system and a type of hydraulic controller. Discuss the advantages and disadvantages of hydraulic control.
15. Sketch and describe a high lift boiler safety valve and state its advantages over the ordinary spring-loaded type.
16. Explain what is meant by "Open Loop" control. Give an example of such a system and state its disadvantages.
17. Sketch and describe a flash evaporator and explain the principle of flash evaporation.
18. Explain the principle of heat sensing by bi-metallic strips and describe a fire detection and alarm system that uses bi-metal strips as detectors.
19. Describe a method of piston cooling in a diesel engine. What cooling medium would you expect to find in the pistons of an opposed piston engine?
20. Briefly describe the sequence of events following the movement of the control lever from stop to an ahead position on the bridge for a remote controlled diesel engine.
21. Of what is brine composed for use in a refrigeration system? In the event of a deficiency how could an emergency brine be made? Could sea water be used to replace the loss?
22. What is a pyrometer? Explain the principle of operation of a resistance type of temperature measuring device.
23. State clearly the difference between an impulse and a reaction turbine, explaining the principle of each and the basic differences in their construction. Where is a dummy piston fitted and why?
24. What is meant by data logging? Describe a data-logging system suitable for an automated engine room.
25. Sketch and describe any type of hydraulically operated steering gear and explain how it is controlled from the bridge.
26. Describe a tanker's automatic cargo control system. Include in your description necessary alarms, controls, instrumentation and fail safe arrangements.
27. Trace the path of the fuel oil in a diesel engine installation from its storage in the double bottom tanks to the cylinders.
28. Explain how the stalling of a diesel engine is prevented in a remote controlled installation such as bridge control.

29. Describe a device for measuring the brake or shaft power of the main engine on board ship. What is the difference between brake power and indicated power?
30. Make a simple sketch and briefly describe the operation of a pneumatic relay, or controller, in a controlled system.
31. Describe the warming through process from cold, of a steam turbine, in preparation to sailing.
32. Describe a smoke detection and CO_2 smothering system for hold compartments which uses one pipe to each compartment.
33. Make a simple sketch and describe a steam boiler suitable for using the waste heat in the exhaust gases from a diesel engine.
34. What are analogue and digital indicators? Explain their advantages and disadvantages.
35. Explain with the aid of a sketch a refrigeration system suitable for use on board ship. Why is the absorption system not used in ships?
36. Describe with the aid of a sketch a ship's draught indicator that gives remote display.
37. Describe with sketches the cycle of operations in a two-stroke diesel engine. Explain different methods of scavenging.
38. Describe a remote control system for a tank valve in an oil tanker and explain how indication of valve position is given.
39. Sketch and describe a thrust block. Explain how the power is transmitted from the engine to the propeller, and how the thrust is transmitted to the ship.
40. Describe with the aid of a simple sketch a "Whessoe" tank gauge and explain how it gives a reading at a position remote from the tank.
41. What type of engine usually drives the emergency generator? Describe the process of starting up an emergency generator and how the load is transferred from the main to the emergency switch-board.
42. What is meant by closed loop control? Illustrate your answer by using a cascade control system.
43. What is meant by a "scavenge fire"? Explain the cause and means of extinguishing. What precautions should be taken to prevent scavenge fires?
44. Describe with the aid of a simple sketch a proportional controller of the hydraulic type. What is meant by

proportional control?

45. Sketch and describe a device incorporating a valve which allows hot water to pass but restricts the passage of steam. Give examples where you could expect one of these to be fitted.
46. Sketch, and fully describe the operation of, an automatic smoke detector.
47. Why are some diesel engines warmed through before starting? Describe the operation.
48. What is the sequence of events following the sensing of a plant fault in an automatic data logged control system? What is a transducer?
49. State the advantages and disadvantages of carbon dioxide, ammonia, and freon 12 as a refrigerant.
50. Describe a CO_2 total flooding system for the machinery spaces of a vessel. Explain how such a system would be operated.

SELECTION OF QUESTIONS ON CALCULATIONS

1. A vessel left port with 300 tonne of fuel in the bunkers, covered a journey of 2,448 nautical miles at a speed of 12 knots and arrived at her destination with 45 tonne of fuel in hand. What speed would have enabled the ship to arrive at her destination with the same reserve of fuel if she started off from port with 233 tonne in the bunkers? What would be the daily consumption of fuel at this new speed?
2. The pitch of a ship's propeller is 5m. It is required to steam a distance of 4 nautical miles by estimation from the engine revolution counter, at an engine speed of 78 rev/min. Assuming a propeller slip of 5% calculate the number of revolutions required to travel this distance, and the time taken.
3. A ship travelled a distance of 800 nautical miles at a speed of 10 knots and burned 100 tonne of fuel on the voyage. How far could this ship travel at a speed of 12 knots on a total consumption of 220 tonne?
4. Explain the meaning of the term "slip."
   A ship's propeller has a pitch of 5.1m; in 24 hours steaming at a propeller speed of 91 rev/min, it was found that the vessel logged 331 nautical miles, find the percentage slip.
5. When the speed of a ship is 11 knots her daily fuel consumption is 44 tonne. Calculate the total fuel required over a voyage of 3,960 nautical miles at a speed of 10 knots.
6. The pitch angle of a ship's propeller, measured at a radius of 2.5m from the centre of the boss, is 21°48′. Calculate the pitch of the blades at this radius.
7. The volume of a mass of gas at a temperature of 0°C and pressure 1,013 millibars is 240cm³. Calculate its volume when its temperature and pressure is 15°C and 1,030 millibars respectively.
8. A ship travelling at 12 knots burns 24 tonne of fuel per day. She left port with 470 tonne of fuel in the bunkers on an

intended voyage of 3,864 nautical miles. After the first three days at sea, she changed her course and proceeded at 15 knots to go to the assistance of a distressed ship, however, after three days on this new course which added 68 nautical miles to the voyage, it was found that assistance was not required. Estimate the speed at which she should complete the voyage to enable her to arrive in port with 67 tonne of fuel to spare.

9. A single-cylinder four-stroke diesel engine has a cylinder diameter of 180mm and a stroke of 250mm. If the mean effective pressure is 4.48 bar when running at 360 rev/min, calculate the indicated power developed.
10. When steaming at 12 knots, a vessel burns 30 tonne of fuel per day. At what speed should she travel in order to cover a voyage of 1,600 nautical miles on a total fuel consumption of 150 tonne?
11. A ship's engine runs at a steady speed of 92 rev/min and the ship travels a distance of 3,700 nautical miles in 10½ days. If the pitch of the propeller is 5.5 metres, calculate the average percentage slip.
12. A ship covers a voyage of 2,400 nautical miles at a speed of 10 knots and consumes 600 tonne of fuel. What would be the probable consumption of fuel over the same distance at a speed of 11 knots?
13. The temperature of a mass of gas of volume 3.8m^3 and pressure 1.5 bar is 25°C. Calculate the final temperature after it has been expanded to 4.5m^3 and 1.2 bar.
14. A ship covers a distance of 410 nautical miles in one day at an engine speed of 1.6 rev/sec. If the pitch of the propeller is 6 metres, find the percentage slip.
15. It is required to steam at 12 knots. If the pitch of the propeller is 4.5 metres and a slip of 5% is assumed, what should be the engine speed in rev/min?
16. The displacement of a ship when light is 6,000 tonne. She takes on board 7,500 tonne of cargo, 300 tonne of stores and 200 tonne of fuel. Find at what speed she should travel in order to do 700 nautical miles on half her bunker fuel, given that the consumption of fuel per day at dead weight displacement is 15,000 tonne and speed 14 knots, is 60 tonne.
17. A six-cylinder two-stroke diesel engine has a cylinder diameter of 685mm and a stroke of 1 ,050mm. Calculate the

total indicated power of the engine when the mean effective pressure in each cylinder is 5.5 bar and the speed 114 rev/min.

18. The designed pitch of the propeller of a new ship is 6.1 metres. Allowing for an estimated 5% slip, calculate the time taken to do the measured mile at a propeller speed of 100 rev/min.
19. A ship on a voyage of 5,500 nautical miles travels at a speed of 12 knots burning 36 tonne of fuel per day. After six days on the voyage, orders are received to proceed to her destination at maximum possible speed. If 900 tonne of fuel are in the bunkers when these orders are received, calculate the maximum speed of the ship so as to arrive in port with 70 tonne reserve and the number of days it will take to reach port.
20. A vessel covers a distance of 75 nautical miles over the ground in 7 hours against a 3 knot current when the propeller is turning at 86 rev/min. If the pitch is 5.5 metres, calculate the actual percentage slip.
21. The speed of a ship is 10 knots when her engines develop 3,000kW. What will be the power of the engines when the speed is 12 knots?
22. A ship commences a voyage of 5,280 nautical miles with 750 tonne of fuel in the bunkers, travelling at 12 knots and burning 30 tonne of fuel per day. After 10 days at sea, 200 tonne of fuel are accidentally lost overboard, at what speed should the ship complete the passage so as to arrive in port with 50 tonne in hand?
23. At 12 noon the engine indicator read 402,500. 24 hours later the reading was 493,100. If the pitch of the propeller is 4.25 metres, calculate the day's run assuming a slip of 9%.
24. A ship burns 40 tonne of fuel per day when her displacement is 16,500 tonne and speed 14.5 knots. Estimate the consumption of fuel for a voyage of 2,040 nautical miles at a speed of 17 knots when the displacement is 13,000 tonne.
25. The pitch angle of the blades of a propeller measured at radii of 0.5, 1.0, 1.5 and 2.0 metres are respectively 40, 29, 24 and 21 degrees. Calculate the pitch of the propeller.
    During a voyage when the propeller was turning at 103 rev/min the actual speed of the ship was 12.8 knots, calculate the percentage slip.

Solutions to Calculations

1. Fuel used for the voyage in the first case.

$$= 300 - 45 = 255 \text{ tonne.}$$

Fuel used for the voyage in the second case.

$$= 233 - 45 = 188 \text{ tonne.}$$

$$\frac{\text{New voyage consumption}}{\text{Old voyage consumption}} = \left\{\frac{\text{New speed}}{\text{Old speed}}\right\}^2$$

$$\frac{188}{255} = \frac{V^2}{12^2}$$

$$\therefore \text{New speed} = \sqrt{\frac{188 \times 144}{255}}$$

$$= 10.3 \text{ knots}$$ Ans.(i)

Number of days on voyage at 10.3 knots,

$$= \frac{2{,}448}{10.3 \times 24}$$

$$\text{Fuel used per day} = \frac{\text{total fuel}}{\text{number of days}}$$

$$= \frac{188 \times 10.3 \times 24}{2{,}448}$$

$$= 19 \text{ tonne per day}$$ Ans. (ii)

2. In one revolution of the propeller, distance travelled by ship

$$= 0.95 \times 5 \text{ metres.}$$

Total distance to travel $= 4 \times 1.852 \times 10^3$ metres

$\therefore$ number of revolutions to be turned by propeller

$$= \frac{4 \times 1.852 \times 10^3}{0.95 \times 5}$$

$$= 1{,}560 \text{ revolutions}$$ Ans. (i)

$$\text{Time taken} = \frac{\text{Distance}}{\text{speed}}$$

$$= \frac{\text{number of revolutions}}{\text{revolutions per minute}}$$

$$= \frac{1{,}560}{78}$$

$$= 20 \text{ minutes}$$ Ans. (ii)

3. $$\frac{\text{New voyage consumption}}{\text{Old voyage consumption}} = \left\{\frac{\text{New speed}}{\text{Old Speed}}\right\}^2 \times \frac{\text{New distance}}{\text{Old distance}}$$

$$\frac{220}{100} = \frac{12^2}{10^2} \; \frac{\text{New distance}}{800}$$

$$\text{New distance} = \frac{220 \times 100 \times 800}{100 \times 144}$$

$$= 1{,}222 \text{ nautical miles} \qquad \text{Ans.}$$

4. See Chapter 9 for explanation of the term 'slip'.

$$\text{Propeller speed [knots]} = \frac{\text{pitch [m]} \times \text{rev/min} \times 60}{10^3 \times 1.852}$$

$$= \frac{5.1 \times 91 \times 60}{10^3 \times 1.852}$$

$$= 15.04 \text{ knots.}$$

$$\text{Ship's speed} = \frac{331}{24}$$

$$= 13.79 \text{ knots}$$

$$\text{per cent slip} = \frac{\text{propeller speed} - \text{ship's speed}}{\text{propeller speed}} \times 100$$

$$= \frac{15.04 - 13.79}{15.04} \times 100$$

$$= 8.312 \text{ per cent} \qquad \text{Ans.}$$

5. $$\frac{\text{New voyage consumption}}{\text{Old voyage consumption}} = \left\{\frac{\text{New speed}}{\text{Old Speed}}\right\}^2 \times \frac{\text{New distance}}{\text{Old distance}}$$

$$\frac{\text{New voyage consumption}}{44} = \frac{10^2}{11^2} \times \frac{3{,}960}{11 \times 24}$$

$$\text{New voyage consumption} = \frac{44 \times 100 \times 3{,}960}{121 \times 11 \times 24}$$

$$= 545.5 \text{ tonne} \qquad \text{Ans.}$$

6. Pitch = 21π x radius x tan of pitch angle
= 2 x π x 2.5 x tan 21° 48'
= 6.282 metres Ans.

7. $$\frac{p_1V_1}{T_1} = \frac{P_2V_2}{T_2}$$

$$\frac{1{,}013 \times 240}{(O + 273)} = \frac{1{,}030 \times V_2}{(15 + 273)}$$

$$V_2 = \frac{1{,}013 \times 240 \times 288}{1{,}030 \times 273}$$

$$= 249\text{cm}^3 \qquad \text{Ans}$$

8. Distance travelled during first three days,

$= 12 \times 24 \times 3 = 864$ nautical miles.

Fuel burned during first three days,

$= 24 \times 3 = 72$ tonne.

Distance travelled during next three days.

$= 15 \times 24 \times 3 = 1{,}080$ nautical miles.

Fuel burned during this time,

$$= \left\{\frac{15}{12}\right\}^3 \times 24 \times 3 = 140.6 \text{ tonne}$$

(note that fuel per day varies as speed^3).

Distance to complete voyage

$= 3{,}864 - 864 - 1{,}080 + 68 = 1{,}988$ nautical miles

Fuel available to complete voyage

$= 470 - 72 - 140.6 - 67 = 190.4$ tonne.

At original speed of 12 knots, consumption to complete voyage of 1,988 nautical miles

$= \text{number of days} \times \text{tonne per day}$

$$= \frac{1{,}988}{12 \times 24} \times 24$$

$= 165.7$ tonne.

$$\frac{\text{New consumption per voyage}}{\text{Old consumption per voyage}} = \left\{\frac{\text{New speed}}{\text{Old speed}}\right\}^2$$

$$\therefore \text{New speed} = \sqrt{\frac{190.4 \times 12^2}{165.7}}$$

$= 12.86$ knots Ans.

9. Indicated power $= p_m ALn$

where $p_m = 4.48 \text{ bar} = 4.48 \times 10^2 \text{kNm}^2$

$A = 0.7854 \times 0.18^2 \text{m}^2$

$L = 0.25\text{m}$

$$n = \text{rev/s} \div 2 = \frac{360}{60 \times 2} = 3$$

$\text{ip} = p_m ALn$

$= 4.48 \times 10^2 \times 0.7854 \times 0.18^2 \times 0.25 \times 3$

$= 8.553\text{kW}$ Ans.

10. $$\frac{\text{New voyage consumption}}{\text{Old voyage consumption}} = \left\{\frac{\text{New speed}}{\text{Old Speed}}\right\}^2 \times \frac{\text{New distance}}{\text{Old distance}}$$

$$\frac{150}{30} = \frac{\text{New speed}^2}{12^2} \times \frac{1{,}600}{12 \times 24}$$

$$\text{New speed} = \sqrt{\frac{150 \times 12 \times 24 \times 144}{1{,}600 \times 30}}$$

$$= 11.39 \text{ knots}$$ Ans.

11. $$\text{Propeller speed} = \frac{\text{pitch} \times \text{rev/min} \times 60}{10^3 \times 1.852}$$

$$= \frac{5.5 \times 92 \times 60}{10^3 \times 1.852}$$

$$= 16.4 \text{ knots.}$$

$$\text{Ship's speed} = \frac{3{,}700}{10.5 \times 24}$$

$$= 14.69 \text{ knots}$$

$$\text{per cent slip} = \frac{\text{propeller speed} - \text{ship's speed}}{\text{propeller speed}} \times 100$$

$$= \frac{16.4 - 14.69}{16.4} \times 100$$

$$= 10.42 \text{ per cent}$$ Ans.

12. $$\frac{\text{New voyage consumption}}{\text{Old voyage consumption}} = \left\{\frac{\text{New speed}}{\text{Old speed}}\right\}^2$$

$$\frac{\text{New voyage consumption}}{600} = \left\{\frac{11}{10}\right\}^2$$

$$\text{New voyage consumption} = 600 \times 1.1^2$$

$$= 726 \text{ tonne}$$ Ans.

13. $$\frac{p_1V_1}{T_1} = \frac{p_2V_2}{T_2}$$

$$\frac{1.5 \times 3.8}{(25 + 273)} = \frac{1.2 \times 4.5}{T_2}$$

$$T_2 = \frac{298 \times 1.2 \times 4.5}{1.5 \times 3.8} = 282.3\text{K}$$

$$= 282.3 - 273 = 9.3^\circ\text{C}$$ Ans.

14. Propeller speed $= \dfrac{\text{pitch x rev/s x 3,600}}{10^3 \times 1.852}$

$= \dfrac{6 \times 1.6 \times 3{,}600}{10^3 \times 1.852}$

= 18.66 knots.

Ship's speed $= \dfrac{410}{24} =$ 17.08 knots.

per cent slip $= \dfrac{18.66 - 17.08}{18.66} \times 100$

= 8.468 per cent Ans.

15. Propeller speed $= \dfrac{\text{pitch x rev/min x 60}}{10^3 \times 1.852}$

Ship's speed = 0.95 x propeller speed

$\therefore\ 12 = \dfrac{0.95 \times 4.5 \times \text{rev/min} \times 60}{10^3 \times 1.852}$

rev/min $= \dfrac{12 \times 10^3 \times 1.852}{0.95 \times 4.5 \times 60}$

= 86.64 rev/min Ans.

16. New displacement = 6,000 + 7,500 + 300 +200

= 14,000 tonne.

Fuel consumption for the voyage is to be 100 tonne.

$\dfrac{\text{New voyage consumption}}{\text{Old voyage consumption}} =$

$$\left\{\frac{\text{New displ.}}{\text{Old displ.}}\right\}^{2/3} \times \left\{\frac{\text{New speed}}{\text{Old speed}}\right\}^{2} \times \frac{\text{New distance}}{\text{Old distance}}$$

$$\frac{100}{60} = \left\{\frac{14{,}000}{15{,}000}\right\}^{2/3} \times \frac{\text{New speed}^2}{14^2} \times \frac{700}{14 \times 24}$$

$$\text{New speed} = \sqrt{\frac{100 \times 14^3 \times 24}{60 \times 700} \times \left\{\frac{1.5}{1.4}\right\}^{2/3}}$$

= 12 . 81 knots Ans .

17. ip per cylinder $= p_m ALn$

where $p_m = 5.5 \times 10^2$ kN/m^2

$A = 0.7854 \times 0.685^2$ m^2

$L = 1.05$m

$n = 114 \div 60 = 1.9$

ip for six cylinders

$= 5.5 \times 10^2 \times 07854 \times 0.685^2 \times 1.05 \times 1.9 \times 6$

$= 2{,}428$kW Ans.

18. $$\text{Speed of propeller} = \frac{\text{pitch x rev/min x 60 knots}}{10^3 \times 1.852}$$

Speed of ship = 0.95 x propeller speed

$$\text{Time [hours]} = \frac{\text{distance [nautical miles]}}{\text{speed [nautical miles per hour]}}$$

$$\therefore \text{Time [min]} = \frac{1 \times 10^3 \times 1.852}{0.95 \times 6.1 \times 100 \times 60} \times 60$$

= 3.196 minutes Ans.

19. Distance travelled in first six days

= 12 x 24 x 6 = 1,728 nautical miles.

Distance remaining to complete voyage

= 5,500 – 1,728 = 3,772 nautical miles.

Fuel available to complete voyage

= 900 – 70 = 830 tonne.

$$\frac{\text{New voyage consumption}}{\text{Old voyage consumption}} = \left\{\frac{\text{New speed}}{\text{Old speed}}\right\}^2 \times \frac{\text{New distance}}{\text{Old distance}}$$

$$\frac{830}{36} = \frac{\text{New speed}^2}{12^2} \times \frac{3{,}772}{12 \times 24}$$

$$\text{New speed} = \sqrt{\frac{830 \times 144 \times 12 \times 24}{36 \times 3{,}772}}$$

= 15.92 knots Ans. (i)

At 15.92 knots,

$$\text{Number of days to complete voyage} = \frac{3{,}772}{15.92 \times 24}$$

= 9.87 days Ans. (ii)

20. $$\text{Propeller speed} = \frac{\text{pitch x rev/min x 60}}{10^3 \times 1.852}$$

$$= \frac{5.5 \times 86 \times 60}{10^3 \times 1.852}$$

$$= 15.33 \text{ knots.}$$

Ship's speed over ground = 75 ÷ 7 = 10.71 knots

Ship's speed through water = 10.71 + 3 = 13.71 knots

$$\text{per cent slip} = \frac{15.33 - 13.71}{15.33} \times 100$$

$$= 10.56 \text{ per cent}$$ Ans.

21. $$\frac{\text{New power}}{\text{Old power}} = \left\{\frac{\text{New speed}}{\text{Old speed}}\right\}^3$$

$$\frac{\text{New power}}{3{,}000} = \left\{\frac{12}{10}\right\}^3$$

$$\text{New power} = 3{,}000 \times 1.2^3$$

$$= 5{,}184\text{kW}$$ Ans.

22. Fuel used during first 10 days

= 30 x 10 = 300 tonne.

Fuel available to complete voyage

= 750 – 300 – 200 – 50 = 200 tonne.

Distance travelled during first 10 days

= 12 x 24 x 10 = 2,880 nautical miles.

Distance remaining to complete voyage

= 5,280 – 2.880 = 2,400 nautical miles.

$$\frac{\text{New voyage consumption}}{\text{Old voyage consumption}} = \left\{\frac{\text{New speed}}{\text{Old Speed}}\right\}^2 \times \frac{\text{New distance}}{\text{Old distance}}$$

$$\frac{200}{300} = \frac{\text{New speed}^2}{12^2} \times \frac{2{,}400}{2{,}880}$$

$$\text{New speed} = \sqrt{\frac{200 \times 144 \times 2{,}880}{300 \times 2{,}400}}$$

$$= 10.73 \text{ knots}$$ Ans.

23. Revolutions turned by propeller in 24 hours

$$= 493{,}100 - 402{,}500 = 90{,}600$$

Distance travelled by propeller

$$= \frac{\text{pitch x revolutions}}{10^3 \text{ x } 1.852} \text{ nautical miles}$$

Distance travelled by ship

$$= \frac{4.25 \text{ x } 90{,}600}{10^3 \text{ x } 1.852} \text{ x } 0.91$$

$$= 189.2 \text{ nautical miles}$$ Ans.

24. $$\frac{\text{New voyage cons.}}{\text{Old voyage cons.}} = \left\{\frac{\text{New displ.}}{\text{Old displ.}}\right\}^{2/3} \text{ x } \left\{\frac{\text{New speed}}{\text{Old speed}}\right\}^2 \text{ x } \frac{\text{New distance}}{\text{Old distance}}$$

$$\frac{\text{New voyage cons.}}{40} = \left\{\frac{13{,}000}{16{,}500}\right\}^{2/3} \text{ x } \left\{\frac{17}{14.5}\right\}^2 \text{ x } \frac{2{,}040}{14.5 \text{ x } 24}$$

New voyage consumption = 274.8 tonne. Ans.

25. Pitch = $2\pi R \tan\theta$.

At 0.5m radius, pitch = 2πx 0.5 x tan 40° = 2.636m

At lm radius, pitch = 2π x 1 x tan 29° = 3.483m

At 1.5m radius, pitch = 2π x 1.5 x tan 24° = 4.197m

At 2m radius, pitch = 2π x 2 x tan 21° = 4.823m

Propeller pitch = mean value

$$= \frac{2.636 + 3.483 + 4.197 + 4.823}{4}$$

$$= 3.785\text{m}$$ Ans. (i)

$$\text{Propeller speed [knots]} = \frac{\text{pitch [m] x rev/min x 60}}{10^3 \text{ x } 1.852}$$

$$= \frac{3.785 \text{ x } 103 \text{ x } 60}{10^3 \text{ x } 1.852} = 12.63 \text{ knots}$$

$$\text{per cent slip} = \frac{\text{propeller speed} - \text{ship's speed}}{\text{propeller speed}} \text{ x } 100$$

$$= \frac{12.63 - 12.8}{12.63} \text{ x } 100$$

$$= -1.346 \text{ per cent}$$ Ans. (ii)

"Minus slip" is due to the ship travelling at a greater speed than the propeller. This is probably due to a strong following current.

INDEX